WORLDS IN THE MAKING

THE EVOLUTION OF THE UNIVERSE

BY

SVANTE ARRHENIUS

DIRECTOR OF THE PHYSICO-CHEMICAL NOBEL
INSTITUTE, STOCKHOLM

TRANSLATED BY
DR. H. BORNS

ILLUSTRATED

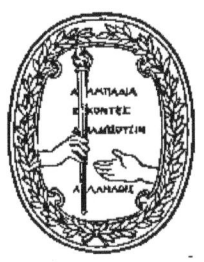

NEW YORK AND LONDON
HARPER & BROTHERS PUBLISHERS
MCMVIII

TABLE OF CONTENTS

I. VOLCANIC PHENOMENA AND EARTHQUAKES . . 1

Destruction caused by volcanism and by earthquakes.—Different kinds of volcanoes.—Vesuvius.—Products of eruption.—Volcanic activity diminishing.—Structure of volcanoes.—Geographical distribution of volcanoes.—Temperature in the interior of the earth.—Significance of water for volcanism.—Composition of the earth's interior.—Geographical distribution of earthquakes.—Fissures in the earth's crust.—Groups of earthquakes.—Waves in the sea and in the air accompanying earthquakes.—Their connection with volcanism.—Systems of fissures.—Seismograms.

II. THE CELESTIAL BODIES, IN PARTICULAR THE EARTH, AS ABODES OF LIVING BEINGS . . . 39

Manifold character of the worlds.—The earth probably at first a ball of gases.—Formation of the earth crust and its rapid cooling.—Balance between heat received and heat lost by radiation.—Life already existing on the earth for a milliard of years.—The waste of solar heat.—Temperature and habitability of the planets.—Heat-preserving influence of the atmosphere.—Significance of carbon dioxide in the atmosphere — Warm and cold geological ages. — Fluctuations in the percentage of carbon dioxide of the air.—Combustion, decay, and growth.—Atmospheric oxygen.—Vegetable life more ancient than animal life.—The atmospheres of planets.—Chances of an improvement in the climate.

III. RADIATION AND CONSTITUTION OF THE SUN . 64

Stability of the solar system.—Losses and possible gains of heat by the sun.—Theses of Mayer and of Helmholtz.—Temperatures of the white, yellow, and reddish stars, and of the sun.—Sun-spots and sun faculæ.—Prominences.—

TABLE OF CONTENTS

Spectra of the parts of the sun.—Temperature of the sun.—The interior of the sun.—Its composition according to the mechanical theory of heat.—The losses of heat by the sun probably covered by the enormous solar energy.

IV. THE RADIATION PRESSURE. 94

Newton's law of gravitation.—Kepler's observation of comets' tails.—The thesis of Euler.—Proof of Maxwell.—The radiation pressure.—Electric charges and condensation.—Comets' tails and radiation pressure.—Constituents and properties of comets' tails.—Weight of the solar corona.—Loss and gain of matter by the sun.—Nature of meteorites.—Electric charge of the sun.—Electrons drawn into the sun.—Magnetic properties of the sun and appearance of the corona.—Constituents of the meteors.—Nebulæ and their heat and light.

V. THE SOLAR DUST IN THE ATMOSPHERE. POLAR LIGHTS AND THE VARIATIONS OF TERRESTRIAL MAGNETISM 118

The supply of dust from the sun rather insignificant.—Polarization of the light of the sky.—The upper clouds.—Different kinds of auroræ.—Their connection with the corona of the sun.—Polar lights and sun-spots.—Periodicity of polar lights.—Polar lights and magnetic disturbances.—Velocity of solar dust.—Fixation of atmospheric nitrogen.—The Zodiacal Light.

VI. END OF THE SUN.—ORIGIN OF NEBULÆ . . . 148

The extinction of the sun.—Collision between two celestial bodies.—The new star in Perseus.—Formation of nebulæ.—The appearance of nebulæ.—The nebulæ catch wandering meteors and comets.—The ring nebula in Lyra.—Variable stars.—Eta in Argus.—Mira Ceti.—Lyra and Algol stars.—Evolution of the stars.

VII. THE NEBULAR AND THE SOLAR STATES . . 191

The energy of the universe.—The entropy of the universe.—The entropy increases in the suns, but decreases in the nebulæ.—Temperature and constitution of the nebulæ.—Schuster's calculations of the condition of a celestial body consisting of gases.—Action of the loss of heat on nebulæ

TABLE OF CONTENTS

and on suns.—Development of a rotating nebula into a planetary system.—The hypothesis of Kant-Laplace.—Objections to it.—The views of Chamberlin and Moulton.—The radiation pressure balances the effect of Newtonian gravitation.—The emission of gases from the nebulæ balances the waste of heat characteristic to the solar systems.

VIII. THE SPREADING OF LIFE THROUGH THE UNIVERSE 212

Stability of the species. — Theory of mutation. — Spontaneous generation.—Bathybius. — Panspermia.—The standpoints of Richter, Ferdinand Cohn, and Lord Kelvin.—The radiation pressure enables spores to escape.—The effect of strong sunlight and of cold on the germinating power.—Transport of spores through the atmosphere into universal space and through it to other planets.—General conclusions.

EXPLANATION OF ABBREVIATIONS, ETC.

The temperatures are stated in degrees centigrade (° C.), either on the Celsius scale, on which the freezing-point of water is 0°, or on the absolute scale, whose zero lies 273 degrees below the freezing-point of water, at $-273°$ C. The equivalent temperatures on the Fahrenheit scale (freezing-point of water 32° F.) are added in brackets (° F.).

1 metre (m.) = 10 decimetres (dm.) = 100 centimetres (cm.) = 1000 millimetres (mm.) = 3.28 ft.; 1 kilometre (km.) = 1000 metres (m.) = 1.6 miles ; 1 mile = 0.62 kilometres (km.).

Light travels in vacuo at the rate of 300,000 km. (nearly 200,000 miles) per second.

ILLUSTRATIONS

FIG.		PAGE
1.	VESUVIUS, AS SEEN FROM THE ISLAND OF NISIDA, IN MODERATE ACTIVITY	2
2.	ERUPTION OF VESUVIUS IN 1882	4
3.	ERUPTION OF VESUVIUS IN 1872	6
4.	PHOTOGRAPH OF VESUVIUS, 1906. CHIEFLY CLOUDS OF ASHES	8
5.	BLOCK LAVA ON MAUNA LOA	10
6.	THE EXCELSIOR GEYSER IN YELLOWSTONE PARK, U. S. A. REMNANT OF POWERFUL VOLCANIC ACTIVITY IN THE TERTIARY AGE	11
7.	MATO TEPEE IN WYOMING, U. S. A. TYPICAL VOLCANIC "NECK"	12
8.	CLEFTS FILLED WITH LAVA AND VOLCANIC CONE OF ASHES, TOROWHEAP CAÑON, PLATEAU OF COLORADO	13
9.	THE KILAUEA CRATER ON HAWAII	15
10.	CHIEF EARTHQUAKE CENTRES, ACCORDING TO THE BRITISH ASSOCIATION COMMITTEE	22
11.	CLEFTS IN VALENTIA STREET, SAN FRANCISCO, AFTER THE EARTHQUAKE OF 1906	25
12.	SAND CRATERS AND FISSURES, PRODUCED BY THE CORINTH EARTHQUAKE OF 1861. IN THE WATER, BRANCHES OF FLOODED TREES	27
13.	EARTHQUAKE LINES IN LOWER AUSTRIA	30
14.	LIBERTY BUILDING OF LELAND STANFORD JUNIOR UNIVERSITY, IN CALIFORNIA, AFTER THE EARTHQUAKE OF 1906	32
15.	EARTHQUAKE LINES IN THE TYRRHENIAN DEPRESSION	34

ILLUSTRATIONS

FIG.		PAGE
16.	SEISMOGRAM RECORDED AT SHIDE, ISLE OF WIGHT, ON AUGUST 31, 1898	35
17.	PHOTOGRAPH OF THE SURFACE OF THE MOON, IN THE VICINITY OF THE CRATER OF COPERNICUS	62
18.	SUN-SPOT GROUP AND GRANULATION OF THE SUN	71
19.	PART OF THE SOLAR SPECTRUM OF JANUARY 3, 1872	75
20.	METALLIC PROMINENCES IN VORTEX MOTION	76
21.	FOUNTAIN-LIKE METALLIC PROMINENCES	76
22.	QUIET PROMINENCES OF SMOKE-COLUMN TYPE	77
23.	QUIET PROMINENCES, SHAPE OF A TREE	77
24.	DIAGRAM ILLUSTRATING THE DIFFERENCES IN THE SPECTRA OF SUN-SPOTS AND OF THE PHOTOSPHERE	78
25.	SPECTRUM OF A SUN-SPOT, THE CENTRAL BAND BETWEEN THE TWO PORTIONS OF THE PHOTOSPHERES PECTRUM	78
26.	THE GREAT SUN-SPOT OF OCTOBER 9, 1903	79
27.	THE GREAT SUN-SPOT OF OCTOBER 9, 1903	80
28.	THE GREAT SUN-SPOT OF OCTOBER 9, 1903	81
29.	THE GREAT SUN-SPOT OF OCTOBER 9, 1903	82
30.	PHOTOGRAPH OF THE SOLAR CORONA OF 1900	83
31.	PHOTOGRAPH OF THE SOLAR CORONA OF 1870	84
32.	PHOTOGRAPH OF THE SOLAR CORONA OF 1898	85
33.	PHOTOGRAPH OF ROERDAM'S COMET (1893 II.), SUGGESTING SEVERAL STRONG NUCLEI IN THE TAIL	100
34.	PHOTOGRAPH OF SWIFT'S COMET (1892 I.)	101
35.	DONATI'S COMET AT ITS GREATEST BRILLIANCY IN 1858	102
36.	IMITATION OF COMETS' TAILS	104
37.	GRANULAR CHONDRUM FROM THE METEORITE OF SEXES. ENLARGEMENT 1 : 70	109
38.	ARCH-SHAPED AURORÆ BOREALIS, OBSERVED BY NORDENSKIOLD DURING THE WINTERING OF THE VEGA IN BERING STRAIT, 1879	124
39.	AURORA BOREALIS, WITH RADIAL STREAMERS	125
40.	AURORA WITH CORONA, OBSERVED BY GYLLENSKIÖLD ON SPITZBERGEN, 1883	126

ILLUSTRATIONS

FIG.		PAGE
41.	POLAR-LIGHT DRAPERIES, OBSERVED IN FINNMARKEN, NORTHERN NORWAY	127
42.	CURVE OF MAGNETIC DECLINATION AT KEW, NEAR LONDON, ON NOVEMBER 15 AND 16, 1905	138
43.	CURVE OF HORIZONTAL INTENSITY AT KEW ON NOVEMBER 15 AND 16, 1905	139
44.	ZODIACAL LIGHT IN THE TROPICS	146
45.	SPECTRUM OF NOVA AURIGÆ, 1892	154
46.	DIAGRAM INDICATING THE CONSEQUENCES OF A COLLISION BETWEEN TWO EXTINCT SUNS	157
47.	SPIRAL NEBULA IN THE CANES VENATICI	159
48.	SPIRAL NEBULA IN THE TRIANGLE	161
49.	THE GREAT NEBULA IN ANDROMEDA	163
50.	RING-SHAPED NEBULA IN LYRA	164
51.	CENTRAL PORTION OF THE GREAT NEBULA IN ORION	165
52.	NEBULAR STRIÆ IN THE STARS OF THE PLEIADES	167
53.	NEBULAR STRIÆ IN THE SWAN	169
54.	NEBULA AND STAR RIFT IN THE SWAN, IN THE MILKY WAY	171
55.	GREAT NEBULA NEAR RHO, IN OPHIUCHUS	172
56.	STAR CLUSTER IN HERCULES	173
57.	STAR CLUSTER IN PEGASUS	175
58.	CONE-SHAPED STAR CLUSTER IN GEMINI	176
59.	COMPARISON OF SPECTRA OF STARS OF CLASSES 2, 3, 4	185
60.	COMPARISON OF SPECTRA OF STARS OF CLASSES 2, 3, 4	186

PREFACE

WHEN, more than six years ago, I was writing my *Treatise of Cosmic Physics*, I found myself confronted with great difficulties. The views then held would not explain many phenomena, and they failed in particular in cosmogonic problems. The radiation pressure of light, which had not, so far, been heeded, seemed to give me the key to the elucidation of many obscure problems, and I made a large use of this force in dealing with those phenomena in my treatise.

The explanations which I tentatively offered could, of course, not claim to stand in all their detail; yet the scientific world received them with unusual interest and benevolence. Thus encouraged, I tried to solve more of the numerous important problems, and in the present volume I have added some further sections to the complex of explanatory arguments concerning the evolution of the Universe. The foundation to these explanations was laid in a memoir which I presented to the Academy of Sciences at Stockholm in 1900. The memoir was soon afterwards printed in the *Physikalische Zeitschrift*, and the subject was further developed in my *Treatise of Cosmic Physics*.

It will be objected, and not without justification, that scientific theses should first be discussed and approved of in competent circles before they are placed before the public. It cannot be denied that, if this condition were

PREFACE

to be fulfilled, most of the suggestions on cosmogony that have been published would never have been sent to the compositors; nor do I deny that the labor spent upon their publication might have been employed for some better purpose. But several years have elapsed since my first attempts in this direction were communicated to scientists. My suggestions have met with a favorable reception, and I have, during these years, had ample opportunity carefully to re-examine and to amend my explanations. I therefore feel justified in submitting my views to a larger circle of readers.

The problem of the evolution of the Universe has always excited the profound interest of thinking men. And it will, without doubt, remain the most eminent among all the questions which do not have any direct, practical bearing. Different ages have arrived at different solutions to this great problem. Each of these solutions reflected the stand-point of the natural philosophers of its time. Let me hope that the considerations which I offer will be worthy of the grand progress in physics and chemistry that has marked the close of the nineteenth and the opening of the twentieth century.

Before the indestructibility of energy was understood, cosmogony merely dealt with the question how matter could have been arranged in such a manner as to give rise to the actual worlds. The most remarkable conception of this kind we find in Herschel's suggestion of the evolution of stellar nebulæ, and in the thesis of Laplace concerning the formation of the solar system out of the universal nebula. Observations more and more tend to confirm Herschel's view. The thesis of Laplace, for a long time eulogized as the flower of cosmogonic speculations, has more and more had to be modified. If we attempt, with Kant, to conceive how wonderfully

PREFACE

organized stellar systems could originate from absolute chaos, we shall have to admit that we are attacking a problem which is insoluble in that shape. There is a contradiction in those very attempts to explain the origin of the Universe in its totality, as Stallo[1] emphasizes: "The only question to which a series of phenomena gives legitimate rise relates to their filiation and interdependence." I have, therefore, only endeavored to show how nebulæ may originate from suns and suns from nebulæ; and I assume that this change has always been proceeding as it is now.

The recognition of the indestructibility of energy seemed to accentuate the difficulties of the cosmogonic problems. The theses of Mayer and of Helmholtz, on the manner in which the Sun replenishes its losses of heat, have had to be abandoned. My explanation is based upon chemical reactions in the interior of the Sun in accordance with the second law of thermodynamics. The theory of the "degradation" of energy appeared to introduce a still greater difficulty. That theory seems to lead to the inevitable conclusion that the Universe is tending towards the state which Clausius has designated as "*Wärme Tod*" (heat death), when all the energy of the Universe will uniformly be distributed through space in the shape of movements of the smallest particles. That would imply an absolutely inconceivable end of the development of the Universe. The way out of this difficulty which I propose comes to this: the energy is "degraded" in bodies which are in the solar state, and the energy is "elevated," raised to a higher level, in bodies which are in the nebular state.

Finally, I wish to touch upon one cosmogonical ques-

---

[1] Stallo: *Concepts and Theories of Modern Physics.* London, 1900, p. 276.

PREFACE

tion which has recently become more actual than it used to be. Some kind of "spontaneous generation," origination of life from inorganic matter, had been acquiesced in. But just as the dreams of a spontaneous generation of energy—i. e., of a *perpetuum mobile*—have been dispelled by the negative results of all experiments in that direction, just in the same way we shall have to give up the idea of a spontaneous generation of life after all the repeated disappointments in this field of investigation. As Helmholtz[1] says, in his popular lecture on the growth of the planetary system (1871): "It seems to me a perfectly just scientific procedure, if we, after the failure of all our attempts to produce organisms from lifeless matter, put the question, whether life has had a beginning at all, or whether it is not as old as matter, and whether seeds have not been carried from one planet to another and have developed everywhere where they have fallen on a fertile soil."

This hypothesis is called the hypothesis of panspermia, which I have modified by combining it with the thesis of the radiation pressure.

My guiding principle in this exposition of cosmogonic problems has been the conviction that the Universe in its essence has always been what it is now. Matter, energy, and life have only varied as to shape and position in space.

THE AUTHOR.

STOCKHOLM, December, 1907.

[1] Helmholtz, *Populäre Wissenschaftliche Vorträge*. Braunschweig, 1876, vol. iii., p. 101.

WORLDS IN THE MAKING

I

VOLCANIC PHENOMENA AND EARTHQUAKES

The Interior of the Earth

THE disasters which have recently befallen the flourishing settlements near Vesuvius and in California have once more directed the attention of mankind to the terrific forces which manifest themselves by volcanic eruptions and earthquakes. The losses of life which have been caused in these two last instances are, however, insignificant by comparison with those which various previous catastrophes of this kind have produced. The most violent volcanic eruption of modern times is no doubt that of August 26 and 27, 1883, by which two-thirds of the island of Krakatoa, 33 square kilometres (13 square miles) in area, situated in the East Indian Archipelago, were blown into the air. Although this island was itself uninhabited, some 40,000 people perished on that occasion, chiefly by the ocean wave which followed the eruption and which caused disastrous inundations in the district. Still more terrible was the destruction wrought by the Calabrian earthquake of February and March, 1783, which consisted of several earthquake waves. The large town of Messina was destroyed on February 5th, and the number of people killed by this event has been estimated at 100,000. The same region, especially Calabria, has, moreover, frequently been visited by disastrous earthquakes—again in 1905 and 1907. Another catastrophe upon which history dwells, owing to the loss of life (not less than 90,000), was the destruction of the capital of Portugal on November 1, 1755. Two-thirds of the human lives which this earthquake claimed were destroyed by a wave 5 m. in height rushing in from the sea.

Vesuvius is undoubtedly the best studied of all volcanoes. During the splendor of Rome this mountain was quite peaceful—known as an extinct volcanic cone so far as history could be traced back. On the extraordinarily fertile soil about it had arisen big colonies of such wealth that the district was called Great Greece (Græcia

Fig. 1.—Vesuvius, as seen from the Island of Nisida, in moderate activity

Magna). Then came, in the year 79 A.D., the devastating eruption which destroyed, among others, the towns of Herculaneum and Pompeii. The volumes of gas, rushing forth with extreme violence from the interior of the earth, pushed aside a large part of the volcanic cone whose remnant is now called Monte Somma, and the falling masses of ashes, mixed with streams of lava,

VOLCANIC PHENOMENA AND EARTHQUAKES

built up the new Vesuvius. This mountain has repeatedly changed its appearance during later eruptions, and was provided with a new cone of ashes in the year 1906. The outbreak of the year 79 was succeeded by new eruptions in the years 203, 472, 512, 685, 993, 1036, 1139, 1500, 1631, and 1660, at quite irregular intervals. Since that time Vesuvius has been in almost uninterrupted activity, mostly, however, of a harmless kind, so that only the cloud of smoke over its crater indicated that the internal glow was not yet extinguished. Very violent eruptions took place in the years 1794, 1822, 1872, and 1906.

Other volcanoes behave quite differently from these violent volcanoes, and do hardly any noteworthy damage. Among these is the crater-island of Stromboli, situated between Sicily and Calabria. This volcano has been in continuous activity for thousands of years. Its eruptions succeed one another at intervals ranging from one minute to twenty minutes, and its fire serves the sailors as a natural light-house. The force of this volcano is, of course, unequal at different periods. In the summer of 1906 it is said to have been in unusually violent activity. Very quiet, as a rule, are the eruptions of the great volcanoes on Hawaii.

Foremost among the substances which are ejected from volcanoes is water vapor. The cloud floating above the crater is, for this reason, the surest criterion of the activity of the volcano. Violent eruptions drive the masses of steam up into the air to heights of 8 km. (5 miles), as the illustrations (Figs. 1 to 4) will show.

The height of the cloud may be judged from the height of Vesuvius, 1300 metres (nearly 4300 ft.) above sea-level. The illustration on page 4 (Fig. 2) is a reproduction of a drawing by Poulett Scrope, representing the

Vesuvius eruption of the year 1822. There seems to have been no wind on this day; the masses of steam formed a cloud of a regular shape which reminds us of a pine-tree. According to the description of Plinius, the cloud noticed at the eruption of Vesuvius in the year 79

Fig. 2.—Eruption of Vesuvius in 1882. (After a contemporaneous drawing by Poulett Scrope)

must have been of the same kind. When the air is not so calm the cloud assumes a more irregular shape (Fig. 3). Clouds which rise to such elevations as we have spoken of are distinguished by strong electric charges. The

VOLCANIC PHENOMENA AND EARTHQUAKES

vivid flashes of lightning which shoot out of the black clouds add to the terror of the awful spectacle. The rain which pours down from this cloud is often mixed with ashes and is as black as ink. The ashes have a color which varies between light-gray, yellow-gray, brown, and almost black, and they consist of minute spherules of lava ejected by the force of the gases and rapidly congealed by contact with the air. Larger drops of lava harden to volcanic sand—the so-called "lapilli" (that is, little stones), or to "bombs," which are often furrowed by the resistance offered by the air, and turn pear-shaped. These solid products, as a rule, cause the greatest damage due to volcanic eruptions. In the year 1906 the weight of these falling masses (Fig. 4) crushed in the roofs of houses. A layer of ashes 7 m. (23 ft.) in thickness buried Pompeii under a protective crust which had covered it up to days of modern excavations. The fine ashes and the muddy rain clung like a mould of plaster to the dead bodies. The mud hardened afterwards into a kind of cement, and as the decomposition products of the dead bodies were washed away, the moulds have provided us with faithful casts of the objects that had once been embedded in them. When the ashes fall into the sea, a layer of volcanic tuffa is formed in a similar manner, which enshrines the animals of the sea and algæ. Of this kind is the soil of the Campagna Felice, near Naples. Larger lumps of solid stones with innumerable bubbles of gases float as pumice-stone on the sea, and are gradually ground down into volcanic sand by the action of the waves. The floating pumice-stone has sometimes become dangerous or, at any rate, an obstacle to shipping, through its large masses; that was, at least, the case with the Krakatoa eruption of 1883.

Among the gases which are ejected in addition to water vapor, carbonic acid should be mentioned in the first instance; also vapors of sulphur and sulphuretted hydrogen, hydrochloric acid, and chloride of ammonium, as well as the chlorides of iron and copper, boric acid, and other substances. A large portion of these bodies is precipitated on the edges of the volcano, owing to

Fig. 3.—Eruption of Vesuvius in 1872. (After a photograph.)

the sudden cooling of the gases. The more volatile constituents, such as carbonic acid, sulphuretted hydrogen, and hydrochloric acid, may spread over large areas, and destroy all living beings by their heat and poison. It was these gases, for example, which caused the awful devastation at St. Pierre, where 30,000 human lives were destroyed on May 8, 1902, by the eruption of Mont Pelée. The ejection of hydrogen gas, which, on emerging from

VOLCANIC PHENOMENA AND EARTHQUAKES

the lava, is burned to water by the oxygen of the air, has been observed in the crater of Kilauea.

The ashes of the volcanoes are sometimes carried to vast distances by the air currents—e.g., from the western coast of South America to the Antilles; from Iceland to Norway and Sweden; from Vesuvius (1906) to Holstein. Best known in this respect is the eruption of the Krakatoa, which drove the fine ashes up to an elevation of 30 km. (18 miles). The finest particles of these ashes were slowly carried by the winds to all parts of the earth, where they caused, during the following two years, the magnificent sunrises and sunsets which were spoken of as "the red glows." This glow was also observed in Europe after the eruption of Mont Pelée. The dust of Krakatoa further supplied the material for the so-called "luminous clouds of the night," which were seen in the years 1883 to 1892 floating at an elevation of about 80 km. (50 miles), and hence illuminated by the light of the sun long after sunset.

The crater of Kilauea, on the high volcano of Mauna Loa, in Hawaii—this volcano is about of the same height as Mont Blanc—has excited special interest. The crater forms a large lake of lava having an area of about 12 sq. km. (nearly 5 sq. miles), which, however, varies considerably with time. The lava boiling at red glow is constantly emitting masses of gas under slight explosions, spurting out fiery fountains to a height of 20 m. (65 ft.) into the air. Here and there lava flows out from crevices in the wall of the crater down the slope of the mountain, until the surface of the lake of lava has descended below these cracks. As a rule, this lava is of a thin fluid consistency, and it spreads, therefore, rather uniformly over large areas. Of a similar kind are also the floods of lava which are sometimes poured over thousands of square kilometres

Fig. 4.—Photograph of Vesuvius, 1906. Chiefly clouds of ashes

on Iceland. The so-called Laki eruption of the year 1783 was of a specially grand nature. Though occurring in an uninhabited district, it did a great amount of damage. In the more ancient geological periods, especially in the Tertiary age, similar sheets of lava of vast extensions have been spread over England and Scotland (more than 100,000 sq. km., roughly, 40,000 sq. miles); over Deccan, in India, 400,000 sq. kms. (150,000 sq. miles), up to heights of 2000 m. (6500 ft.); and over Wyoming, Yellowstone Park, Nevada, Utah, Oregon, and other districts of the United States, as well as over British Columbia.

In other cases the slowly ejected lava is charged with large volumes of gases, which escape when the lava

VOLCANIC PHENOMENA AND EARTHQUAKES

congeals and burst it up into rough, unequal blocks, forming the so-called block lava (Fig. 5). The streams of lava can likewise produce terrible devastation when they descend into inhabited districts; on account of their slow motion, they rarely cause loss of life, however.

Where the volcanic activity gradually lessens or ceases, we can still trace it by the exhalations of gas and the springs of warm water which we find in many districts where, during the Tertiary age, powerful volcanoes were ejecting their streams of lava. To this class belong the famous geysers of Iceland, of Yellowstone Park (Fig. 6), and of New Zealand; also the hot springs of Bohemia, so highly valued therapeutically (e. g., the Karlsbad Sprudel); the Fumaroli of Italy, Greece, and other countries, exhaling water vapor; the Mofettæ, with their exhalations of carbonic acid (of frequent occurrence in the district of the Eifel and on both sides of the middle Rhine, in the Dogs Grotto near Naples, and in the Valley of Death in Java); the Solfatara, exhaling vapors of sulphur—sulphuretted hydrogen and sulphur dioxide (they are found near Naples on the Phlegræan Fields and in Greece); as well as many of the so-called mud volcanoes, which eject mud, salt water, and gases (as a rule, carbonic acid and hydrocarbons)—for example, the mud volcanoes near Parma and Modena, in Italy, and those near Kronstadt, in Transylvania.

The extinct volcanoes, of which some, like the Aconcagua, 6970 m. (22,870 ft.), in South America, and the Kilimanjaro, in Africa, 6010 m. (19,750 ft.), rank among the highest mountains, are exposed to a rapid destruction by the rain, because they consist largely of loose materials—volcanic ashes with interposed layers of lava. Where these lava streams expand gradually, they protect the ground underneath from erosion by water, and

Fig. 5.—Block lava on Mauna Loa

in this way proper cuts are formed on the edges of the lava streams, passing through the old volcano and through the sedimentary strata at deeper levels.

The old volcano of Monte Venda, near Padua, affords an interesting example of this type. We can observe there how the sedimentary limestone has been changed by the lava, which was flowing over it, into marble to a depth of about 1 m. (3 ft.) Sometimes the limestone which is lying over the lava has also undergone the same transformation, which would indicate that lava has not only been flowing above the edge of the crater, but has also forced itself out on the sides through the fissures

Fig. 6.—The Excelsior Geyser in Yellowstone Park, U.S.A. Remnant of powerful volcanic activity in the Tertiary age

between two layers of limestone. Massive subterranean lava streams of this kind are found in the so-called lakkolithes of Utah and in the Caucasus. There the superior layers have been forced upward by the lava pressing from below; the lava froze, however, before it reached the surface of the earth, where it might have formed a volcano. Quite a number of granites, the so-called batholithes, chiefly occurring in Norway, Scotland, and Java, are of similar origin. Occasionally it is only the core of congealed lava that has remained of the whole volcano. These cores, which originally filled the pipe of the crater, are frequent in Scotland and in North America, where they are designated "necks" (Fig. 7).

Fig. 7.—Mato Tepee in Wyoming, U. S. A. Typical volcanic "Neck"

The so-called cañons of the Colorado Plateau, with their almost vertical walls, are the results of the erosive action of rivers. A drawing by Dutton shows a wall of this kind more than 800 m. (2600 ft.) in height, through four fissures of which lava streams have forced their way up to the surface (Fig. 8). Over one of these fissures a small cone of volcanic ashes is still visible, while the cones which probably overtopped the three other fissures have been washed away, so that the veins end in small "necks." Evidently a very fluid lava—strong percentages of magnesia and of oxide of iron render the lava more fluid than an admixture of silicic acid, and the fluidity is further increased by the presence of water—has been forced into the fissures which were already present, and has reached the surface of the earth before it froze. The

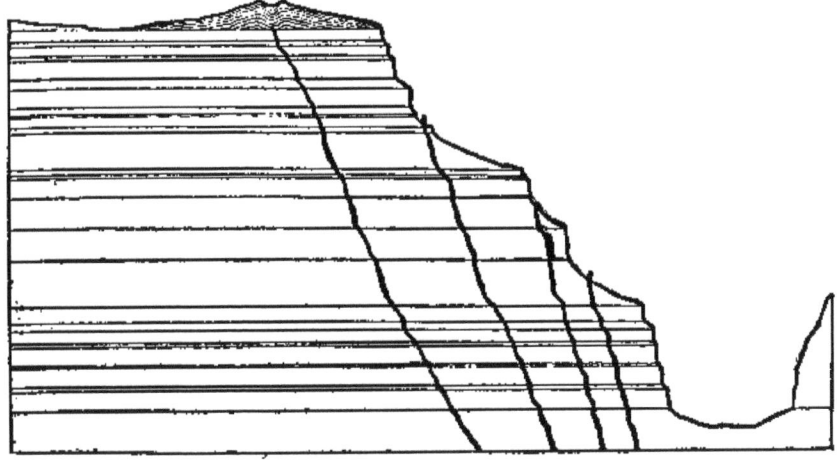

Fig. 8.—Clefts filled with lava and volcanic cone of ashes, Torowheap Cañon, Plateau of Colorado. Diagram.

driving force behind them must have been pretty strong; else the lava streams could not have attained the necessary velocity of flow.

When the Krakatoa was blown into the air in 1883 half of the volcano remained behind. This half clearly shows the section of the cone of ashes, which has been but very slightly affected by the destructive action of the water. We find there in the central part the light-colored stopper of lava in the volcano pipe, and issuing from it more light-colored beds of lava, between which darker strata of ashes can be seen.

The distribution of volcanoes over the surface of the earth is marked by striking regularities. Almost all the volcanoes are situated near the shores of the sea. A few are found in the interior of East Africa; but they are, at any rate, near the Great Lakes of the equatorial regions. The few volcanoes which are supposed to be situated in Central Asia must be regarded as doubtful. We miss, however, volcanoes on some sea-coasts, as in Australia and along the long coast-lines of the Northern Arctic Ocean to the north of Asia, Europe, and America. Volcanoes occur only where great cracks occur in the crust of the earth along the sea-coast. Where such fissures are found, but where the sea or large inland lake basins are not near—as, for instance, in the Austrian Alps—we do not meet with any volcanoes; such districts are, however, renowned for their earthquakes.

Since ancient ages the belief has been entertained that the molten masses of the interior of the earth find an outlet through the volcanoes.. Attempts have been made to estimate the depth of the hearths of volcanoes, but very different values have been deduced. Thus, the hearth under the volcano of Monte Nuovo, which was thrown up in the year 1538 on the Phlegræan Fields, near Naples, has been credited with depths varying from 1.3 km. to 60 km. (1 mile to 40 miles); for the Krakatoa, estimates of more than 50 km. (30 miles) have been made. All these calculations are rather aimless; for the volcanoes are probably situated on folds of the earth-crust, through which the fluid mass (the magma) rushes forth in wedges from the interior of the earth, and it will presumably be very difficult to say where the hearth of magma ends and where the volcanic pipe commences. The Kilauea gives the visitor the impression that he is standing over an opening in the crust of the earth, through which the molten mass rushes forth directly from the interior of the earth. (Fig. 9.)

As regards the earth-crust, we know from observations in bore-holes made in different parts of the world that the temperature increases rather rapidly with the depth,

VOLCANIC PHENOMENA AND EARTHQUAKES

on an average by about thirty degrees Cent. per kilometre (about 1.6° F. per 100 feet). It must be remarked, however, that the depth of our deepest bore-holes hardly exceeds 2 km. (Paruchowitz, in Silesia, 2003 m., or 6570 ft.; Schladebach, near Merseburg, Prussian Saxony, 1720 m.). If the temperature should go on increasing at the rate of 30 degrees Cent. for each further kilometre, the temperature at a depth of 40 kilometres should attain degrees at which all the common rocks would melt. But the melting-point certainly rises at the same time as the pressure. The importance of this circumstance was,

Fig. 9.—The Kilauea Crater on Hawaii

however, much exaggerated when it was believed that for this reason the interior of the earth might possibly be solid. Tammann has shown by direct experiments that the temperature of fusion only rises up to a certain pressure, and that it begins to decrease again on a further increase of pressure. The depths indicated above are therefore not quite correct. If we assume, however, that other kinds of rock behave like diabase—the melting-point of which, according to the determinations of Barus, rises by 1° Cent. for each 40 atmospheres of pressure corresponding to a depth of 155 m.—we should conclude that the solid crust of the earth could not have a greater thickness than 50 or 60 km. (40 miles). At greater depths we should therefore penetrate into the fused mass. On account of its smaller density the silicic acid will be concentrated in the upper strata of the molten mass, while the basic portions of the magma, which are richer in iron oxide, will collect in the lower strata, owing to their greater density.

This magma we have to picture to ourselves as an extremely viscid liquid resembling asphalt. The experiments of Day and Allen show that rods, supported at their ends, of $30 \times 2 \times 1$ mm. of different minerals, like the feldspars microcline and albite, could retain their shape for three hours without curving noticeably, although their temperature was about a hundred degrees above their melting-point, and although they appeared completely fused, or, more correctly, completely vitrified when taken out of the furnace. These molten silicates behave very differently from other liquids like water and mercury, with which we are more accustomed to deal.

The motion and diffusion in the magma, and especially in the very viscous and sluggish acid portions of the upper strata, will therefore be exceedingly small, and the magma will behave almost like a solid body, like the minerals of the experiments of Day and Allen. The magmas of volcanoes like Etna, Vesuvius, and Pantellaria may, therefore, have quite different compositions, as we should conclude from their lavas without our being forced to believe, with Stübel, that these three hearths of volcanoes are completely separated, though not far removed from one another. In the lava of Vesuvius a temperature of 1000 or 1100 degrees has been found at the lower extremity of the stream. From the occurrence in the lava of certain crystals like leucite and olivin, which we have reason to assume must have been formed before the lava left the crater, it has been concluded that the lava temperature cannot have been higher than 1400 degrees before it left the volcanic pipe.

It would, however, be erroneous to deduce from the temperature of the lava of Vesuvius that the hearth of the volcano must be situated at a depth of approximately 50 kilometres. Most likely its depth is much smaller, perhaps not even 10 kilometres. For there, as everywhere where volcanoes occur, the crust of the earth is strongly furrowed, and the magma will just at the spots where we find volcanoes come much nearer to the surface of the earth than elsewhere.

The importance of water for the formation of volcanoes probably lies in the fact that, in the neighborhood of cracks under the bottom of the sea, the water penetrates down to considerable depths. When the water reaches a stratum of a temperature of 365 degrees —the so-called critical temperature of water—it can no longer remain in the liquid state. That would not prevent, however, its penetrating still farther into the depths, in spite of its gaseous condition. As soon as the vapor comes in contact with magma, it will eagerly be absorbed by the magma. The reason is that water of a temperature of more than 300 degrees is a stronger acid than silicic acid; the latter is therefore expelled by it from its compounds, the silicates, which form the main constituents of the magma. The higher the temperature, the greater the power of the magma to absorb water. Owing to this absorption the magma swells and becomes at the same time more fluid. The magma is therefore pressed out by the action of a pressure which is analogous to the osmotic pressure by virtue of which water penetrates through a membrane into a solution of sugar or salt. This pressure may become equivalent to thousands of atmospheres, and this very pressure would raise the magma up the volcanic pipe even to a height of 6000 m. (20,000 feet) above the sea-level. As the magma is ascending in the volcanic pipe it is slowly cooled, and its capacity for binding water diminishes with falling temperature. The water will hence escape under violent ebullition, tearing drops and larger lumps of lava with it, which fall down again as ashes or pumice-stone. After the lava has flown out of the crater and is slowly cooling, it continues to give off water, breaking up under the formation of block lava (see Fig. 5). If, on the other hand, the lava in the crater of the volcano is comparatively at rest, as in Kilauea, the water will escape more slowly; owing to the long-continued contact of the surface layer of lava with the air, little water will remain in it, the water being, so to say, removed by aeration, and the lava streams will therefore, when congealing, form more smooth surfaces.

In some cases volcanoes have been proved (Stübel and Branco) not to be in connection with any fractures in the crust of the earth. That holds, for instance, for several volcanoes of the early Tertiary age in Swabia. We may imagine that the pressure produced by the swelling of the magma became so powerful as to be able to break through the earth-crust at thinner spots, even in the absence of previous fissures.

VOLCANIC PHENOMENA AND EARTHQUAKES

If, in our consideration, we follow the magma farther into the depths, we shall not find any reason for assuming that the temperature will not rise farther towards the interior of the earth. At depths of 300 or 400 km. (250 miles) the temperature must finally attain degrees such that no substance will be able to exist in any other state than the gaseous. Within this layer the interior of the earth must, therefore, be gaseous. From our knowledge of the behavior of gases at high temperatures and pressures, we may safely conclude that the gases in the central portions of the earth will behave almost like an extremely viscid magma. In certain respects they may probably be compared to solid bodies; their compressibility, in particular, will be very small.

We might think that we could not possibly learn anything concerning the condition of those strata. Earthquakes have, however, supplied us with a little information. Such gaseous masses must fill by far the greatest part of the earth, and they must have a very high specific gravity; for the average density of the earth is 5.52, and the outer strata, the ocean and the masses of the surface which are known to us, have smaller densities. The ordinary rocks possess a density ranging from 2.5 to 3. It must, therefore, be assumed that the materials of the innermost portions of the earth must be metallic, and Wiechert, in particular, has advocated this view. Iron will presumably form the chief constituent of this gas of the central earth. Spectrum analysis teaches us that iron is a very important constituent of the sun. We know, further, that the metallic portions of the meteorites consist essentially of iron; and finally terrestrial magnetism indicates that there must be large masses of iron in the interior of the earth. We have also reason to believe that the native iron occurring in nature—*e. g.*, the well-known iron of Ovifak, in Greenland—is of volcanic origin. The materials in the gaseous interior of the earth will, owing to their high density, behave in chemical and physical respects like liquids. As substances like iron will, also at very high temperatures, have a far higher specific gravity than their oxides, and these again have a higher gravity than their silicates, we have to assume that the gases in the core of the earth will almost exclusively be metallic, that the outer portions of the core will contain essentially oxides, and those farther out again mostly silicates.

The fused magma will, on penetrating in the shape of batholithes into the upper layers, probably be divided into two portions, of which one, the lighter and gaseous, will contain water and substances soluble in it; while the other, heavier portion, will essentially consist of silicates with a lower percentage of water. The more fluid portion, richer in water, will be secreted in the higher layers, will penetrate into the surrounding sedimentary strata, especially into their fissures, and will fill them with large crystals, often of metallurgical value—*e.g.*, of the ores of tin, copper, and other metals, while the water will slowly evaporate through the superposed strata. The more viscid and sluggish mass of silicates, on the other hand, will congeal, thanks to its great viscosity, to glass, or, when the cooling is very slow, to small crystals.

We now turn to earthquakes. No country has been absolutely spared by earthquakes. In the districts bounding upon the Baltic, and especially in northern Russia, they have, however, been of a quite harmless type. The reason is that the earth-crust there has been lying undisturbed for long geological epochs and has never been fractured. The comparatively severe earthquake which shook the west coast of Sweden on October 23, 1904, to

VOLCANIC PHENOMENA AND EARTHQUAKES

an unusually heavy degree, without, however, causing any noteworthy damage (a few chimneys were knocked over), was caused by a fault of relatively pronounced character for those districts in the Skager-Rack—a continuation of the deepest fold in the bottom of the North Sea, the so-called Norwegian Trough, which runs parallel to the Norwegian coast. In Germany, the Vogtland and the districts on both sides of the middle Rhine have frequently been visited by earthquakes. Of other European countries, Switzerland, Spain, Italy, and the Balkan Peninsula, as well as the Karst districts of Austria, have often suffered from earthquakes.

According to the committee appointed by the British Association for the investigation of earthquakes—a committee which has contributed a great deal to our knowledge of these great natural phenomena—earthquakes of some importance emanate from certain centres which have been indicated on the subjoined map (Fig. 10). The most important among these regions comprises Farther India, the Sunda Isles, New Guinea, and Northern Australia; it is marked on the map by the letter F. From this district have emanated in the six-year period 1899–1904 no fewer than 249 earthquakes, which have been recorded in many observatories far removed from one another. This earthquake centre F is closely related to the one marked E, in Japan, from which 189 earthquakes have proceeded. Next to this comes the extensive district K with 174 earthquakes, comprising the most important folds in the crust of the Old World, including the mountain chains from the Alps to the Himalaya. This district is interesting, because it has been disturbed by a great many earthquakes, although it is almost entirely situated on the Continent. After that we have the dis-districts A, B, C, with 125, 98, and 95 earthquakes. They are situated near lines of fracture in the earth-crust along the American coast of the Pacific Ocean and the Caribbean Sea. District D, with 78 earthquakes, is similarly situated. The three last-mentioned districts, B, C, D, as well as G, between Madagascar and India, with 85 earthquakes, all seem to be surpassed by the district H in the eastern Atlantic, with its 107 earthquakes. These latter are, however, relatively feeble, and we owe their accurate records probably to the circumstances that a great many earthquake observatories are situated within the immediate surroundings of this district. The same may be said of the district I, or Newfoundland, which is not characterized by many earthquakes, and of the district J, between Iceland and Spitzbergen, with 31 and 19

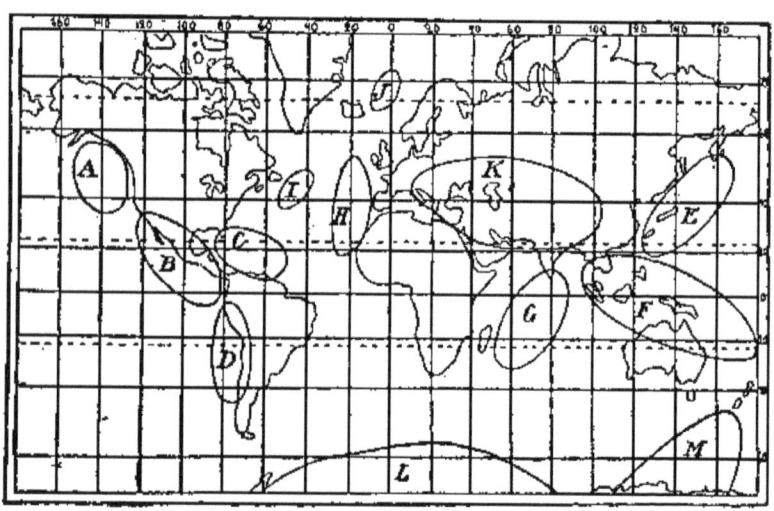

Fig. 10.—Chief earthquake centres, according to the British Association Committee

earthquakes respectively. The last on the list used to be the district L, situated about the South Pole, with only eight earthquakes. This small number is probably mere-

VOLCANIC PHENOMENA AND EARTHQUAKES

ly due to the want of observatories in those parts of the earth. Another district, M, has finally been added, which extends to the southwest from New Zealand. No fewer than 75 intense earthquakes were recorded between March 14 and November 23, 1903, by the Discovery Expedition, in 70° southern latitude and 178° eastern longitude. Earthquakes commonly occur in swarms or groups. Thus, more than 2000 shocks were counted on Hawaii in March, 1868. During the earthquakes which devastated the district of Phokis, in Greece, in 1870-73, shocks succeeded one another for a long time at intervals of three seconds. During the whole period of three and a half years about half a million shocks were counted, and, further, a quarter of a million subterranean reports which were not accompanied by noticeable concussions. Yet of all these shocks only about 300 did noteworthy damage, and only 35 were considered worth being reported in the newspapers. The concussion of October 23, 1904, belonged to a group which lasted from October 10 to October 28, and in which numerous small tremors were noticed, especially on October 24 and 25. The earthquake of San Francisco commenced on April 18, 1906, at 5 hrs. 12 min. 6 sec. A.M. (Pacific Ocean time), and ended at 5 hrs. 13 min. 11 sec., lasting therefore 1 minute and 5 seconds. Twelve smaller shocks succeeded in the following hour. Before 6 hrs. 52 min. P.M., nineteen further concussions were counted, and various smaller shocks succeeded in the following days.

With such groups of earthquakes weaker tremors usually precede the violent destructive shocks and give a warning. Unfortunately this is not always so, and no warning was given by the earthquakes which destroyed Lisbon in 1755 and Caracas in 1812, nor by those which devastated Agram in 1880, nor, finally, in the case of the San Fran-

WORLDS IN THE MAKING

cisco disaster. A not very severe earthquake without feebler precursors befell Ischia in 1881, while the violent catastrophe which devastated this magnificent island in 1883 was heralded by several warnings. As in San Francisco and Chili in 1906, less violent concussions generally succeed the destructive shocks. Earthquakes like that of Lisbon in 1755, consisting of a single shock, are very rare.

The violent concussions often produce large fissures in the ground. Such were noticed in several places at San Francisco. One of the largest fissures known, that of Midori, in Japan, was caused by the earthquake of October 20, 1891. It left a displacement of the ground ranging up to 6 m. (20 ft.) in the vertical and 4 m. (13 ft.) in the horizontal direction. This crack had a length of not less than 65 km. (40 miles). Extensive fissures were also formed by the earthquakes of Calabria, in 1783, at Monte San Angelo, and in the sandstones of the Bálpakrám Plateau in India, in 1897. In mountainous districts falls of rock are a frequent consequence of the formation of fissures and earthquakes. A large number of rocks fell in the neighborhood of Delphi during the Phokian earthquake. On January 25, 1348, an earthquake sent down a large portion of Mount Dobratsch (in the Alps of Villach, in Carinthia, which is now much frequented by tourists) and buried two towns and seventeen villages. The earthquake of April 18, 1906, in California started from a crack which extends from the mouth of Alder Creek, near Point Arena, running parallel with the coast-line mostly inland, then entering the sea near San Francisco, and turning again inland between Santa Cruz and San José, finally proceeding *via* Chittenden up to Mount Pinos, a distance of about 600 km. (400 miles), in the direction of N. 35° W. to S. 35° E. Along this

VOLCANIC PHENOMENA AND EARTHQUAKES

Fig. 11.—Clefts in Valentia Street, San Francisco, after the earthquake of 1906

crack the two masses of the earth have been displaced so that the ground situated to the southwest of the fissure has been moved by about 3 m. (10 ft.), and in some spots even by 6 m. (20 ft.) towards the northwest. In some localities in Sonoma and Mendocino counties the southwestern part has been raised, but nowhere by more than 1.2 m. (4 ft.). This is the longest crack which has ever been noticed in connection with an earthquake.

The earthquake over, the ground does not always return to its original position, but remains in a more or less wavy condition. This can most easily be observed in districts where streets or railways cross the ground. It is reported, for instance, that the track of the tramway-lines in Market Street, the chief thoroughfare of San Francisco, formed large wavelike curves after the earthquake.

As a consequence of the displacements in the interior of the earth and of the formation of fissures, river courses are changed, springs become exhausted, and new springs arise. That was the case, for instance, in California in 1906. The ground water often rushes out with considerable violence, tearing with it sand and mud and stones, and piling them up, occasionally forming little craters (Fig. 12). Extensive floods may also be caused on such occasions. By such a flood the ancient Olympia was submerged under a layer of river sand which for some time preserved from destruction the ancient Greek masterpieces of art—among them the famous statue of Hermes. The floods afterwards receded, and the treasures of ancient Olympia could be excavated.

Like the natural water channels and arteries in the interior of the earth, water mains are displaced by the concussions. The direct damage caused by the floods is often less important than the damage due to the impossibility of extinguishing the fires which follow the destruction of the buildings. It was the fires that did most of the enormous material damage in the destruction of San Francisco.

Still greater devastation is wrought by the ocean waves thrown up by earthquakes. We have already referred to the flood of Lisbon in 1755, which was felt on the western coast of Norway and Sweden. Another wave, in 1510, devoured 109 mosques and 1070 houses in Constantinople. Another wave, again, invaded Kamaïshi, in Japan, on June 15, 1896, swept away 7600 houses and killed 27,000 people.

We have repeatedly alluded to the disastrous floodwave of Krakatoa of 1883. This wave traversed the whole of the Indian Ocean, passing to the Cape of Good Hope and Cape Horn, and travelled round half the globe afterwards. Even more remarkable was the aerial wave, which spread like an explosion wave.

While the most violent cannonades are rarely heard for more than 150 km. (95 miles) in a single case at a distance of 270 km. (170 miles) the eruption of

VOLCANIC PHENOMENA AND EARTHQUAKES

Krakatoa was heard at Alice Springs, at a distance of 3600 kilometres, and on the island of Rodriguez, at almost 4800 km. (3000 miles). The barographs of the meteorological stations first marked a sudden rise and then a decided sinking of the air pressure, succeeded by a few smaller fluctuations. These air pulses were repeated in some places as many as seven times. We may therefore assume that the aerial wave passed these places three

Fig. 12.—Sand craters and fissures, produced by the Corinth earthquake of 1861. In the water, branches of flooded trees

times in the one direction, and three times in the other, travelling round the earth. The velocity of propagation of this wave was 314.2 m. (1030 ft.) per second, corresponding to a temperature of —27° Cent. (17° F.) which prevails at an altitude of about 8 km. (5 miles) above the earth's surface, at which altitude this wave may have travelled.

Within the last decade a peculiar phenomenon (leading to what is designated variation of latitudes) has been studied. The poles of the axis of the earth appear to move in a very irregular curve about their mean axis. The movement is exceedingly small. The deviation of the North Pole from its mean position does not amount to more than 10 m. (about 33 ft.). It has been believed that these motions of the North Pole are subject to sudden fluctuations after unusually violent earthquakes, especially when such concussions follow at rapid intervals. That would give us, perhaps more than any other observation, an idea of the force of earthquakes, since they would appear to be able to disturb the equilibrium of the whole mass of our globe.

A severely felt effect of earthquakes, though most people perhaps pay little attention to it, is the destruction of submarine cables. The gutta-percha sheaths of cables are frequently found in a fused condition, suggesting volcanic eruptions under the bottom of the sea. We take care now to avoid earthquake centres in laying telegraphic cables. Their positions have been ascertained by the most modern investigations (see Fig. 10).

People have always been inclined to look for a connection between earthquakes and volcanic eruptions. The connection is unquestionable in a large number of violent earthquakes. In order to establish it, the above-mentioned committee of the British Association has compiled the following table of the history of the earthquakes of the Antilles:

1692. Port Royal, Jamaica, destroyed by an earthquake; land sinking into the sea. Eruption on St. Kitts.

VOLCANIC PHENOMENA AND EARTHQUAKES

1718.—Terrible earthquake on St. Vincent, followed by an eruption.
1766-67.—Great shocks in northeastern South America, in Cuba, Jamaica, and the Antilles. Eruption on Santa Lucia.
1797.—Earthquake in Quito, loss of 40,000 lives. Concussions in the Antilles, eruption on Guadeloupe.
1802.—Violent shocks in Antigua. Eruption on Guadeloupe.
1812.—Caracas, capital of Venezuela, totally destroyed by earthquake. Violent shocks in the Southern States of North America, commencing on November 11, 1811. Eruptions on St. Vincent and Guadeloupe.
1835-36.—Violent concussions in Chili and Central America. Eruption on Guadeloupe.
1902.—April 19. Violent shocks, destroying many towns of Central America. Mont Pelée, on Martinique, in activity. Eruption on May 3. Submarine cables break, sea recedes. Renewed violent movements of the sea on May 8, 19, 20. Eruption on St. Vincent, cable destroyed on May 7. Violent eruption of Mont Pelée on May 8. Destruction of St. Pierre. Numerous smaller earthquakes.

This table distinctly marks the restless state of affairs in that part of the earth, and how quiet and safe matters are comparatively in old Europe, especially in the north. Some parts of Central America are so persistently visited by earthquakes that one of them, Salvador, has been christened "Schaukelmatte." It is not saying too much to assert that the earth is there incessantly trembling. Other districts which are very frequently visited are the Kuriles and Japan, as well as the East Indian islands. In all these countries the crust of the earth has been broken and folded within comparatively recent epochs (chiefly in the Tertiary age) by numerous fissures, and their compression is still going on.

The smaller earthquakes, of which not less than 30,000 are counted in the course of a year, do not stand in any closer relation to volcanic eruptions. This is also the

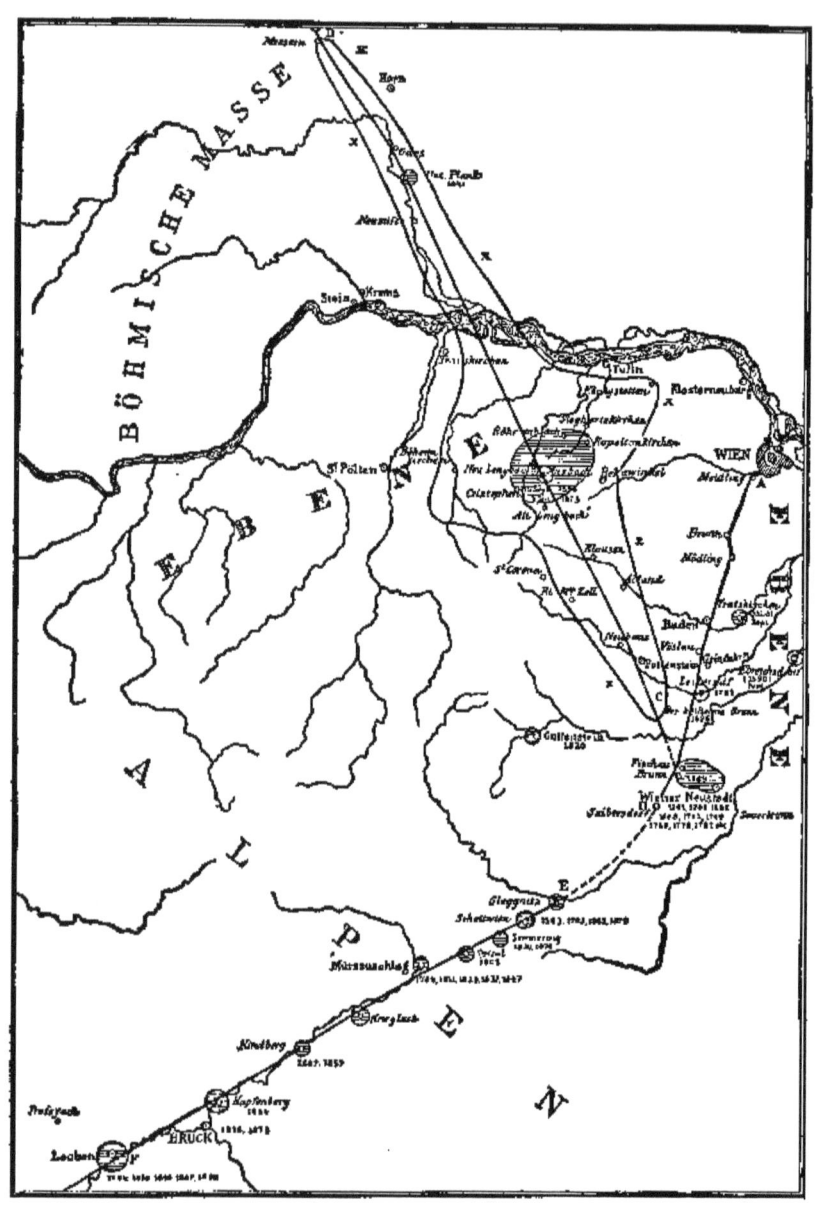

Fig. 13.—Earthquake lines in lower Austria

VOLCANIC PHENOMENA AND EARTHQUAKES

case for a number of large earthquakes, among which we have to count the San Francisco earthquake.

It is assured with good reason that earthquakes are often produced at the bottom of the sea, where there is a strong slope, by slips of sedimentary strata which have been washed down from the land into the sea in the course of centuries. Milne believes that the seaquake of Kamaïshi of June 15, 1896, was of this character. Concussions may even be promoted by the different loading of the earth resulting from the fluctuations in the pressure of the air above it.

Smaller, though occasionally rather violent, earthquakes are not infrequent in the neighborhood of Vienna. On the map (Fig. 13) we see three lines. The line A B is called the thermal line, because along it a number of hot springs, the thermæ of Meidling, Baden, Vöslau, etc., are located, which are highly valued; the other line B C is called the Kamp line, because it is traversed by the river Kamp; and the third B F is called the Mürz line, after the river Mürz. The main railway-track between Vienna and Bruck follows the valleys of A B and E F.

These lines, which probably correspond to large fissures in the earth-crust, are known as sources of numerous earthquakes. The district about Wiener Neustadt, where the three lines intersect, is often shaken by violent earthquakes; some of their dates have been marked on the map.

The curve which is indicated by the letters X X on the map marks the outlines of an earthquake which started on January 3, 1873, from both sides of the Kamp line. It is striking to see how the earthquake spread in the loose ground of the plain between St. Pölten and Tulln, while the masses of rock situated to the northwest and southeast formed obstacles to the propagation of the earthquake waves.

WORLDS IN THE MAKING

Similar conclusions have been deduced from the study of the spreading of the waves which destroyed Charleston, South Carolina, in 1886. Twenty-seven lives were destroyed by this shock. It was the most terrible earthquake that ever visited the United States before the year 1906. In the Charleston concussion the Alleghany Mountains proved a powerful bar against the

Fig. 14.—Library building of Leland Stanford Junior University, in California, after the earthquake of 1906. The photograph shows the great strength of iron structures in comparison to the strength of brickwork. The effect of the earthquake on wooden structures can be seen in Fig. 11

VOLCANIC PHENOMENA AND EARTHQUAKES

further propagation of the shocks, which all the more easily travelled in the loose soil of the Mississippi Valley. In San Francisco, likewise, the worst devastation fell upon those parts of the town which had been built upon the loose, partly made ground in the neighborhood of the harbor, while the buildings erected on the famous mountain ridges of San Francisco suffered comparatively little damage, in so far as they were not reached by the destructive fires. As regards the destructive effects of the earthquake in San Francisco, the building-ground of that city has been divided into four classes (the first is the safest, the last the most unsafe)—namely: 1. Rocky soil. 2. Valleys situated between rocks and filled up by nature in the course of time. 3. Sand-dunes. 4. Soil created by artificial filling up. This latter soil "behaved like a semiliquid jelly in a dish," according to the report of the Earthquake Commission.

For similar reasons the sky-scrapers, constructed of steel on deep foundations, stood firmest. After them came brick houses, with well-joined and cemented walls on deep foundations. The weakness of wooden houses proved mainly due to the poor connection of the beams, a defect which might easily be remedied. The superiority of the steel structure will be apparent from the illustrations (Figs. 11 and 14).

The spots situated just over the crack, of which we spoke on page 25, suffered the most serious damage. Next to them, devastation befell especially localities which, like Santa Rosa, San José, and Palo Alto with Leland Stanford Junior University, are situated on the loose soil of the valley, whose deepest portions are covered by the bay of San Francisco. The splendidly endowed California University, in Berkeley, and the famous Lick Observatory, both erected on rocky ground, fortunately escaped without any notable damage.

The map sketch (Fig. 15) by Suess represents the earthquake lines of Sicily and Calabria. These districts have, as mentioned before, been devastated by severe earthquakes, of which the most terrible occurred in the year 1783, and again in 1905 and 1907. They have also been the scene of many smaller concussions.

The bottom of the Tyrrhenian Sea — between Italy, Sicily, and Sardinia—has been lowered in rather recent ages

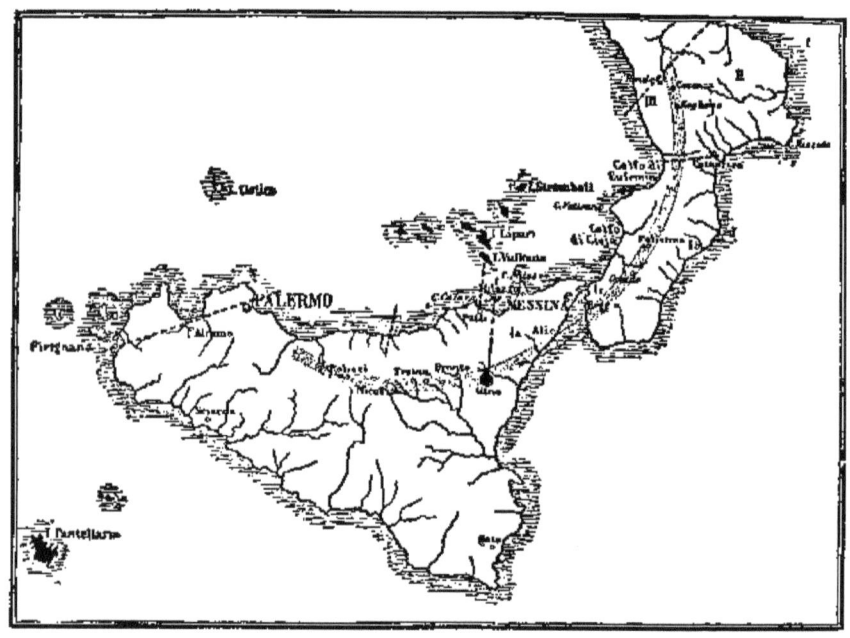

Fig. 15.—Earthquake lines in the Tyrrhenian depression

and is still sinking. We notice on the map five dotted lines, corresponding to cracks in the crust of the earth. These lines would intersect in the volcanic district of the Lipari Islands. We further see a dotted circular arc corresponding to a fissure which is regarded as the source of the Calabrian earthquakes of 1783, 1905, and 1907. The earth-

VOLCANIC PHENOMENA AND EARTHQUAKES

crust behaved somewhat after the manner of a windowpane which was burst by a heavy impact from a point corresponding to the Island of Lipari. From this point radiate lines of fracture, and fragments have been broken off from the earth-crust by arc-shaped cracks, The volcano Etna is situated on the intersection of the radial and circular fissures.

In recognition of the high practical importance of earthquake observations, seismological stations have in recent days been erected in many localities. At these observatories the earthquakes are recorded by pendulums whose styles draw lines on tapes of paper moved by clock-work. As long as the earth is quiet the drawn line is straight. When earthquakes set in, the line passes into a wavy curve. As long as the movement of the paper is slow, the curve merely looks like a widened straight line. The subjoined illustration (Fig. 16) represents a seismogram taken at the station of Shide, on the Isle of Wight, on August 31, 1898. The earthquake recorded originated in the Centre G, in the Indian Ocean. The origin has been deduced from the

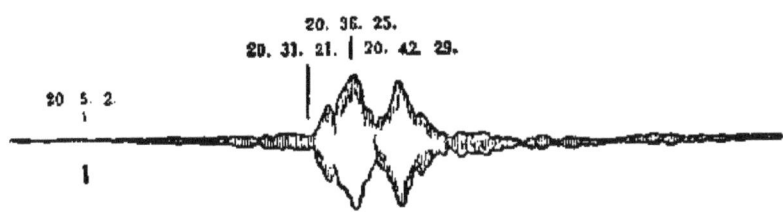

Fig. 16.—Seismogram recorded at Shide, Isle of Wight, on August 31, 1898

moments of arrival of the different waves at different stations. We notice on the seismogram a faint widening of the straight line at 20 hrs. 5 min. 2 sec. (8 hrs. 5 min. 2 sec. P.M.). The amplitude of the oscillations then began to widen, and the heaviest concussions were noticed at 20 hrs. 36 min. 25 sec., and 20 hrs. 42 min. 49 sec., after which the amplitudes slowly decreased with smaller shocks. The first shock of 20 hrs. 5 min. 2 sec. is called the preliminary tremor. This tremor passes through the interior of the earth at a velocity of propagation of 9.2 km. (5¾ miles) per second. It would require twenty-three minutes to pass through the earth along a diameter. The tremor is very feeble, which is ascribed to the extraordinarily great friction characteristic of the strongly heated gases which are confined in the interior of the earth. The principal violent shock at 20 hrs. 36 min. 25 sec. was caused by a wave travelling through the solid crust of the earth. The intensity of this shock is much less impaired than that of the just-mentioned tremor, and it travels with the smaller velocity of about 3.4 km. (2.1 miles) along the earth's surface.

The velocity of propagation of concussion pulses has been calculated for a mountain of quartz, in which it would be 3.6 km. (2.2 miles) per second, very nearly the same as the last-mentioned figure. We should expect this, since the firm crust of the earth consists essentially of solid silicates —*i.e.*, compounds of quartz endowed with similar properties.

Measured at small distances from the origin, the velocity of propagation of the wave appears smaller, and the first preliminary tremor is frequently not observed. The velocity may be diminished to 2 km. (1¼ miles) per second. The reason is that the pulse partly describes a curve in the more solid portions of the crust, and partly passes through looser strata, through which the wave travels at a much slower rate than in firm ground; for instance, at 1.2 km. through loose sandstones, at 1.4 km. through the water of the ocean, and at 0.3 km. through loose sand.

VOLCANIC PHENOMENA AND EARTHQUAKES

We recognize that it should be possible to calculate the distance between the point of observation and the origin of the earthquake from the data relating to the arrivals of the first preliminary tremor and of the principal shock of maximum amplitude. The violent shock is sometimes repeated after a certain time, though with decreased intensity. It has often been observed that this secondary, less violent, shock seems to have travelled all round the earth *via* the longest road between the origin and the point of observation, just like one portion of the aerial waves in the eruption of Krakatoa (compare page 27). The velocity of propagation of this secondary shock is the same as that of the principal shock.

Milne has deduced from his observations that, when the line joining the origin of the earthquake and the point of observation does not at its lowest level descend deeper than 50 km. below the surface of the earth, the pulse will travel undivided through the solid crust of the earth. For this reason we estimate the thickness of the solid crust at 50 km. The value is in almost perfect agreement with the one which we had (on page 16) derived from the increase of temperature with greater depths. It should further be mentioned, perhaps, that the density of the earth in the vicinity has been determined from pendulum observation, and that this density seems to be rather variable down to the depths of 50 or 60 km., but to become more uniform at greater depths. These 50 or 60 km. (31 or 37 miles) would belong to the solid crust of the earth.

The movement of earthquake shocks through the earth thus teaches us that the solid earth-crust cannot be very thick, and that the core of the earth is probably gaseous. The similar conclusions, to which these various considerations had led us, may therefore come very near the truth. A careful study of seismograms may, we hope, help us to learn more about the central portions of the earth, which at first sight appear to be absolutely inaccessible to scientific research.

II

THE CELESTIAL BODIES, IN PARTICULAR THE EARTH, AS ABODES OF ORGANISMS

THERE is no more elevating spectacle than to contemplate the sky with its thousands of stars on a clear night. When we send our thoughts to those lights glittering in infinite distance, the question forces itself upon us, whether there are not out there planets like our own that will sustain organic life. How little interest do we take in a barren island of the Arctic Circle, on which not a single plant will grow, compared to an island in the tropics which is teeming with life in its most wonderful variety! The unknown worlds occupy our minds much more when we may fancy them inhabited than when we have to regard them as dead masses floating about in space.

We have to ask ourselves similar questions with regard to our own little planet, the earth. Was it always covered with verdure, or was it once sterile and barren? And if that be so, what are the conditions under which the earth can fulfil its actual part of harboring organic life? That "the earth was without form" in the beginning is unquestionable. It does not matter whether we assume that it was once all through an incandescent liquid, which may be the most probable assumption, or that it was, as Lockyer and Moulton think, formed by the accumulation of meteoric stones which became incandescent when arrested in their motion.

We have seen that the earth probably consists of a mass of gas encased within a shell which is solid on the outside and remains a viscid liquid on the inner side. We presume with good reason that the earth was originally a mass of gas separated from the sun, which is still in the same state. By radiation into cold space the sphere of gas which, on the whole, would behave as our sun does now, would gradually lose its high temperature, and finally a solid crust could form on its surface. Lord Kelvin has calculated that it would not require more than one hundred years before the temperature of this crust would sink to 100°. Supposing, even, that Kelvin's calculations should not quite be confirmed, we may yet maintain that not many thousands of years would have elapsed from the time when the earth assumed its first crust at about 1000° till the age when this temperature had fallen below 100° (212° F.). Living beings certainly could not exist so long, since the albumen of the cells would at once coagulate at the temperature of boiling water, like the white of an egg. Yet it has been reported that some of the hot springs of New Zealand contain algæ, although at a temperature of over 80°. When I went to Yellowstone Park to inquire into the correctness of this statement, I found that the algæ existed only at the edge of the hot springs, where the temperature did not exceed 60° (140° F.). The famous American physiologist Loeb states that we do not meet with algæ in hot springs at temperatures above 55°.

Since, now, the temperature of the earth-crust would much more quickly sink from 100° to 55° than it had fallen from 1000° to 100°, we may imagine that only a few thousands of years may have intervened between the formation of the first crust of the earth and the cooling down to a temperature such as would sustain life. Since that time the temperature has probably never been so low that the

CELESTIAL BODIES AS ABODES OF ORGANISMS

larger portion of the earth's surface would not have been able to support organisms, although there have been several glacial ages in which the arctic districts inaccessible to life must have extended much farther than at present. The ocean will also have been free of ice over much the greatest portion of its surface at all times, and may therefore have been inhabited by organisms in all ages. The interior of the earth cools continually, though slowly, because heat passes from the inner, warmer portions to the other, cooler portions through the crust of the earth.

The earth is able to serve as the abode of living beings because its outer portions are cooled to a suitable temperature (below 55°) by radiation, and because the cooling does not proceed so far that the open sea would continually be frozen over, and that the temperature on the Continent would always remain below freezing-point. We owe this favorable intermediate stage to the fact that the radiation from the sun balances the loss of heat by radiation into space, and that it is capable of maintaining the greater portion of the surface of the earth at a temperature above the freezing-point of water. The temperature conditioning life on a planet is therefore maintained only because, on the one side, light and heat are received by radiation from the sun in sufficient quantities, while on the other side an equivalent radiation of heat takes place into space. If the heat gain and the heat loss were not to balance each other, the term of suitable conditions would not last long. The temperature of the earth-crust could sink in a few hundreds or thousands of years from 1000° to 100°, because when the earth was at this high temperature its radiation into space predominated over the radiation received from the sun. On the other hand, about a hundred million years have passed, according to Joly, since the age when the ocean originated. The temperature of the earth, therefore, required this long space of time in order to cool down from 365° (at which temperature water vapor can first be condensed to liquid water) to its present temperature. The cooling afterwards proceeded at a slower rate, because the difference between the radiations inward and outward was lessened with the diminishing temperature of the earth. Various methods have been applied in estimating these periods. Joly based his estimate on the percentage of salt in the sea and in the rivers. If we calculate how much salt there is in the sea, and how much salt the rivers can supply to it in the course of a year, we arrive at the result that the quantity of salt now stored in the ocean might have been supplied in about a hundred million years.

We arrive at still higher numbers when we calculate the time which must have elapsed during the deposition of all the stratified and sedimentary layers. Sir Archibald Geikie estimates the total thickness of those strata, supposing them to have been undisturbed, at 30,000 m. (nearly 20 miles). He concludes, further, from the examination of more recent strata, that every stratum one metre in thickness must have required from 3000 to 20,000 years for its formation. We should, therefore, have to allow a space of from ninety to six hundred million years for the deposition of all the sedimentary strata. The Finnish geologist Sederholm even fixes the time at a thousand million years.

Another method again starts from the consideration that, while the temperature of the surface of the earth remains fairly steady owing to the heat exchange between solar radiation and terrestrial radiation into space, the interior of the earth must have shrunk with the cooling. How far this shrinkage extends we may estimate from the

CELESTIAL BODIES AS ABODES OF ORGANISMS

formation of the mountain chains which, according to Rudzki, cover 1.6 per cent. of the earth's surface. The earth's radius should consequently have contracted by about 0.8 per cent., corresponding to a cooling through about 300°, which would require two thousand million years.

Quite recently the renowned physical chemist Rutherford has expounded a most original method of estimating the age of minerals. Uranium and thorium are supposed to produce helium by their slow dissociation, and we know how much helium is produced from a certain quantity of uranium or thorium in a year. Now Ramsay has determined the percentage of helium in the uranium mineral fergusonite and in thorianite. Rutherford then calculates the time which would have passed since the formation of these minerals. He demands at least four hundred million years, "for very probably some helium has escaped from the minerals during that time." Although this estimate is very uncertain, it is interesting to find that it leads to an age for the solid earth-crust of the same order of magnitude as the other methods.

During this whole epoch of almost inconceivable length of between one hundred million and two thousand million years, organisms have existed on the surface of the earth and in the sea which do not differ so very much from those now alive. The temperature of the surface may have been higher than it is at present; but the difference cannot be very great, and will amount to 20° Cent. (36° F.) at the highest. The actual mean temperature of the surface of the earth is 16° Cent. (61° F.). It varies from about $-20°$ Cent. ($-4°$ F.) at the North Pole, and $-10°$ Cent. ($+14°$ F.) at the South Pole to 26° Cent. (79° F.) in the tropical zone. The main difference between the temperatures of the earth's surface in the most remote period from which fossils are extant and the actual state rather seems to be that the different zones of the earth are now characterized by unequal temperatures, while in the remote epochs the heat was almost uniformly distributed over the whole earth.

The condition for this prolonged, almost stationary state was that the gain of heat of the earth's surface by radiation from the sun and the loss of heat by radiation into space nearly balanced each other. That the replenishing supply by radiation from an intensely hot body—in our case the sun—is indispensable for the existence of life will be evident to everybody. Not everybody may, however, have considered that the loss of heat into cold space or into colder surroundings is just as indispensable. To some people, indeed, the assumption that the earth as well as the sun should waste the largest portions of their vital heat as radiation into cold space appears so unsatisfactory that they prefer to believe radiation to be confined to radiation between celestial bodies; there is no radiation into space, in their opinion. All the solar heat would thus benefit the planets and the moons in the solar system, and only a vanishing portion of it would fall upon the fixed stars, because their visual angles are so small. If that were really correct, the temperature of the planets would rise at a rapid rate until it became almost equal to that of the sun, and all life would become impossible. We are therefore constrained to admit that "things are best as they are," although the great waste of solar heat certainly weakens the solar energy.

The opinion that all the solar heat radiated into infinite space is wasted, starts moreover from a hypothesis which is not proved, and which is highly improbable—namely, that only an extremely small portion of the sky is covered with celestial bodies. That might certainly be correct if we assumed, as has formerly been done, that the majority of the celestial bodies must be luminous. We do not

CELESTIAL BODIES AS ABODES OF ORGANISMS

possess, however, any reliable knowledge of the number and size of the dark celestial bodies. In order to account for the observed movements of different stars, it has been thought that there must be in the neighborhood of some of them dark stars of enormous size whose masses would surpass the mass of our sun, or, at least, be equal to it. But the largest number of the dark celestial bodies which hide the rays from the stars behind them probably consist of smaller particles, such as we observe in meteors and in comets, and to a large extent of so-called cosmical dust. The observations of later years, by the aid of most powerful instruments, have shown that so-called nebulæ and nebulous stars abound throughout the heavens. In their interior we should probably find accumulations of dark masses.

The light intensity of most of the nebulæ is, moreover, far too weak to permit of their being perceived. We have, therefore, to imagine that there are bodies all through infinite space, and about as numerous as they are in the immediate neighborhood of our solar system. Thus every ray from the sun, of whatever direction, would finally hit upon some celestial body, and nothing would be lost of the solar radiation, nor of the stellar radiation.

As regards the radiation-heat exchange, the earth might be likened to a steam-engine. In order that the steam-engine shall perform useful work, it is necessary not only that the engine be supplied with heat of high temperature from a furnace and a boiler, but also that the engine be able to give its heat up again to a heat reservoir of lower temperature—a condenser or cooler. It is only by transferring heat from a body of higher temperature to another body of lower temperature that the engine can do work. In a similar way no work can be done on the earth, and no life can exist, unless heat be conferred by the intermediation of the earth from a hot body, the sun, to the colder surroundings of universal space—*i. e.*, to the cold celestial bodies in it.

To a certain extent the temperature of the earth's surface, as we shall presently see, is conditioned by the properties of the atmosphere surrounding it, and particularly by the permeability of the latter for the rays of heat.

If the earth did not possess an atmosphere, or if this atmosphere were perfectly diathermal — *i. e.*, pervious to heat radiations—we should be able to calculate the mean temperature of the earth's surface, given the intensity of the solar radiation, from Stefan's law of the dependence of heat radiation on its temperature. Starting from the not improbable assumption that, at a mean distance of the earth from the sun, the solar rays would send 2.5 gramme-calories per minute to a body of cross section of 1 sq. centimetre at right angles to the rays of the sun, Christiansen has calculated the mean temperatures of the surfaces of the various planets. The following table gives his figures, and also the mean distances of the planets from the sun, in units of the mean distance of the earth from the sun, 149.5 million km. (nearly 93 million miles):

Planet	Radius According to See	Mass According to See	Mean distance	Mean temperature	Density according to See
Mercury...	0.311	0.0224	0.39	+ 178°(332°)	0.564
Venus.....	0.955	0.815	0.72	+ 65°	0.936
Earth.....	1	1	1	+ 6.5°	1
Moon......	0.273	0.01228	1	+ 6.5°(105°)	0.604
Mars......	0.53	0.1077	1.52	− 37°	0.729
Jupiter....	11.13	317.7	5.2	− 147°	0.230
Saturn ...	9.35	95.1	9.55	− 180°	0.116
Uranus....	3.35	14.6	19.22	− 207°	0.388
Neptune ..	3.43	17.2	30.12	− 221°	0.429
Sun.......	109.1	332,750.	0	+6200°	0.256

In the case of Mercury, I have added another figure, 332°. Mercury always turns the same side to the sun, and the hottest point of this side would reach a temperature of 397°; its mean temperature, according to my calculation, is 332°, while the other side, turned away from the sun, cannot be at a temperature much above absolute zero, −273°. I have made a similar calculation for the moon, which turns so slowly about its axis (once in twenty-seven days) that the temperature on the side illuminated by the sun remains almost as high (106°) as if the moon were always turning the same face to the sun. The hottest point of this surface would attain a temperature of 150°, while the poles of the moon and that part of the other side which remains longest without illumination can, again, not be much above absolute zero temperature. This estimate is in fair agreement with the measurements made of the lunar radiation and the temperature estimate based upon it. The first measurement of this kind was made by the Earl of Rosse. He ascertained that the moon disk as illuminated by the sun—that is to say, the full moon—would radiate as much heat as a black body of the temperature 110° Cent. (230° F.). A later measurement by the American Very seems to indicate that the hottest point of the moon is at about 180°, which would be 30° higher than my estimate. In the cases of the moon and of Mercury, which do not possess any atmosphere to speak of, this calculation may very fairly agree with the actual state of affairs.

The temperature of the planet Venus would be about 65° Cent. (149° F.) if its atmosphere were perfectly transparent. We know, however, that dense clouds, probably of water drops, are floating in the atmosphere of this planet, preventing us from seeing its land and water surfaces. According to the determinations made by Zöllner and others, Venus would reflect not less than 76 per cent. of the incident light of the sun, and the planet would thus be as white as a snow-ball. The rays of heat are not reflected to the same extent. We may estimate that the portion of heat absorbed by the planet is about half the incident heat. The temperature of Venus will therefore be reduced considerably, but it is partly augmented again by the protective action of this atmosphere. The mean temperature of Venus may, hence, not differ much from the calculated temperature, and may amount to about 40° (104° F.). Under these circumstances the assumption would appear plausible that a very considerable portion of the surface of Venus, and particularly the districts about the poles, would be favorable to organic life.

Passing to the earth, we find that the temperature-reducing influence of the clouds must be strong. They protect about half of the earth's surface (52 per cent.) from solar radiation. But even with a perfectly clear sky, not all the light from the sun really reaches the earth's surface; for finely distributed dust is floating even in the purest air. I have estimated that this dust would probably absorb 17 per cent. of the solar heat. Clouds and dust would therefore together deprive the earth of 34 per cent. of the heat sent to it, which would lead to a reduction of the temperature by about 28°. Dust and the water-bubbles in the clouds also prevent the radiation of heat from the earth, so that the total loss of heat to be charged to clouds and dust will amount to about 20° (36° F.).

It has now been ascertained that the mean temperature of the earth is 16° (61° F.), instead of the calculated 6.5° (43.7° F.). Deducting the 20° due to the influence of dust and clouds, we obtain −14° (7° F.), and the ob-

CELESTIAL BODIES AS ABODES OF ORGANISMS

served temperature would therefore be higher than the calculated by no less than 30° (54° F.). The discrepancy is explained by the heat-protecting action of the gases contained in the atmosphere, to which we shall presently refer (page 51).

There are but few clouds on Mars. This planet is endowed with an atmosphere of extreme transparency, and should therefore have a high temperature. Instead of the temperature of $-37°$ (35° F.), calculated, the mean temperature seems to be $+10°$ ($+50°$ F.). During the winter large white masses, evidently snow, collect on the poles of Mars, which rapidly melt away in spring and change into water that appears dark to us. Sometimes the snow-caps on the poles of Mars disappear entirely during the Mars summer; this never happens on our terrestrial poles. The mean temperature of Mars must therefore be above zero, probably about $+10°$. Organic life may very probably thrive, therefore, on Mars. It is, however, rather sanguine to jump at the conclusion that the so-called canals of Mars prove its being inhabited by intelligent beings. Many people regard the "canals" as optical illusions; Lowell's photographs, however, do not justify this opinion.

As regards the other large planets, the temperatures which we have calculated for them are very low. This calculation is, however, rather illusory, because these planets probably do not possess any solid or liquid surface, but consist altogether of gases. Their densities, at least, point in this direction. In the case of the inner planets, Mars and our moon included, the density is rather less than that of the earth. Mercury stands last among them, with its specific gravity of 0.564. There follows a great drop in the specific gravities of the outer large planets. Saturn, with a density of 0.116, is last in this order;

the densities of the two outermost planets lie somewhat higher—by 0.3 or 0.4 about—but these last data are very uncertain. Yet these figures are of the same order of magnitude as that assumed for the sun—0.25—and we believe that the sun, apart from the small clouds, is wholly a gaseous body. It is therefore probable that the outer planets, including Jupiter, will also be gaseous and be surrounded by dense veils of clouds which prevent our looking down into their interior. That view would contend against the idea that these planets can harbor any living beings. We could rather imagine their moons to be inhabited. If these moons received no heat from their planets, they would assume the above-stated temperatures of their central bodies. Looked at from our moon, the earth appears under a visual angle, 3.7 times as large as that of the sun. As the temperature of the sun has, from its radiation, been estimated at 6200° Cent., or 6500° absolute, the moon would receive as much heat from the earth as from the sun, if the earth had a temperature of about 3100° Cent., or 3380° absolute. When the first clouds of water vapor were being formed in the terrestrial atmosphere, the earth's temperature was about 360°, and the radiation from the earth to the moon only about 1.25-thousandth of that of the sun. The present radiation from the earth does not even attain one-twentieth of this value. It is thus manifest that the radiation from the earth does not play any part in the thermal household of the moon.

The relations would be quite different if the earth had the 11.6 times greater diameter of Jupiter, or the diameter of Saturn, which is 9.3 times greater than its own. The radiation from the earth to the moon would then make up about a sixth or a ninth of the actual solar radiation, taking the temperature of the earth's surface at 360°.

We can easily calculate, further, that Jupiter and Saturn would radiate as much heat against a moon at a distance of 240,000 or 191,000 km. respectively (since the distance of the moon from the earth amounts to 384,000 km.) as the sun sends to Mars — taking the temperature of those planets at 360° Cent. Now we find, near Jupiter as well as near Saturn, moons at the distances of 126,000 and 186,000 km. respectively, which are smaller than those mentioned, and it is not inconceivable that these moons receive from their central bodies sufficient heat to render life possible, provided that they be enveloped by a heat-absorbing atmosphere. The conditions appear to be less favorable for the innermost satellites of Jupiter and Saturn. When their planets are shining at the maximum brilliancy, their light intensity is only a sixth or a ninth of the solar light intensity, which upon these satellites is itself only one-twenty-seventh or one-ninetieth of the intensity on the earth. During the incandescence epoch of these planets their moons will certainly for some time have been suitable for the development of life.

That the atmospheric envelopes limit the heat losses from the planets had been suggested about 1800 by the great French physicist Fourier. His ideas were further developed afterwards by Pouillet and Tyndall. Their theory has been styled the hot-house theory, because they thought that the atmosphere acted after the manner of the glass panes of hot-houses. Glass possesses the property of being transparent to heat rays of small wave lengths belonging to the visible spectrum; but it is not transparent to dark heat rays, such, for instance, as are sent out by a heated furnace or by a hot lump of earth. The heat rays of the sun now are to a large extent of the visible, bright kind. They penetrate through the glass of the hot-house and heat the earth under the glass. The radiation from the earth, on the other hand, is dark and cannot pass back through the glass, which thus stops any losses of heat, just as an overcoat protects the body against too strong a loss of heat by radiation. Langley made an experiment with a box, which he packed with cotton-wool to reduce loss by radiation, and which he provided, on the side turned towards the sun, with a double glass pane. He observed that the temperature rose to 113° (235° F.), while the thermometer only marked 14° or 15° (57° or 59° F.) in the shade. This experiment was conducted on Pike's Peak, in Colorado, at an altitude of 4200 m. (13,800 ft.), on September 9, 1881, at 1 hr. 4 min. P. M., and therefore at a particularly intense solar radiation.

Fourier and Pouillet now thought that the atmosphere of our earth should be endowed with properties resembling those of glass, as regards permeability of heat. Tyndall later proved this assumption to be correct. The chief invisible constitutents of the air which participate in this effect are water vapor, which is always found in a certain quantity in the air, and carbonic acid, also ozone and hydrocarbons. These latter occur in such small quantities that no allowance has been made for them so far in the calculations. Of late, however, we have been supplied with very careful observations on the permeability to heat of carbonic acid and of water vapor. With the help of these data I have calculated that if the atmosphere were deprived of all its carbonic acid— of which it contains only 0.03 per cent. by volume—the temperature of the earth's surface would fall by about 21°. This lowering of the temperature would diminish the amount of water vapor in the atmosphere, and would cause a further almost equally strong fall of temperature. The examples, so far as they go, demonstrate that com-

CELESTIAL BODIES AS ABODES OF ORGANISMS

paratively unimportant variations in the composition of the air have a very great influence. If the quantity of carbonic acid in the air should sink to one-half its present percentage, the temperature would fall by about 4°; a diminution to one-quarter would reduce the temperature by 8°. On the other hand, any doubling of the percentage of carbon dioxide in the air would raise the temperature of the earth's surface by 4°; and if the carbon dioxide were increased fourfold, the temperature would rise by 8°. Further, a diminution of the carbonic acid percentage would accentuate the temperature differences between the different portions of the earth, while an increase in this percentage would tend to equalize the temperature.

The question, however, is whether any such temperature fluctuations have really been observed on the surface of the earth. The geologists would answer: yes. Our historical era was preceded by a period in which the mean temperature was by 2° (3.6 F.) higher than at present. We recognize this from the former distribution of the ordinary hazel-nut and of the water-nut (*Trapa natans*). Fossil nuts of these two species have been found in localities where the plants could not thrive in the present climate. This age, again, was preceded by an age which, we are pretty certain, drove the inhabitants of northern Europe from their old abodes. The glacial age must have been divided into several periods, alternating with intervals of milder climates, the so-called inter-glacial periods. The space of time which is characterized by these glacial periods, when the temperature—according to measurements based upon the study of the spreading of glaciers in the Alps—must have been about 5° (8° F.) lower than now, has been estimated by geologists at not less than 100,000 years.

This epoch was preceded by a warmer age, in which the temperature, to judge from fossilized plants of those days, must at times have been by 8° or 9° (14° or 16° F.) higher than at present, and, moreover, much more uniformly distributed over the whole earth (Eocene). Pronounced fluctuations of this kind in the climate have also occurred in former geological periods.

Are we now justified in supposing that the percentage of carbon dioxide in the air has varied to an extent sufficient to account for the temperature changes? This question has been answered in the affirmative by Högbom, and, in later times, by Stevenson. The actual percentage of carbonic acid in the air is so insignificant that the annual combustion of coal, which has now (1904) risen to about 900 million tons and is rapidly increasing,[1] carries about one-seven-hundredth part of its percentage of carbon dioxide to the atmosphere. Although the sea, by absorbing carbonic acid, acts as a regulator of huge capacity, which takes up about five-sixths of the produced carbonic acid, we yet recognize that the slight percentage of carbonic acid in the atmosphere may by the advances of industry be changed to a noticeable degree in the course of a few centuries. That would imply that there is no real stability in the percentage of carbon dioxide in the air, which is probably subject to considerable fluctuations in the course of time.

Volcanism is the natural process by which the greatest amount of carbonic acid is supplied to the air. Large quantities of gases originating in the interior of the earth are ejected through the craters of the volcanoes. These gases consist mostly of steam and of carbon dioxide, which have been liberated during the slow cooling of the silicates

[1] It amounted in 1890 to 510 million tons; in 1894, to 550; in 1899, to 690; and in 1904, to 890 million tons.

in the interior of the earth. The volcanic phenomena have been of very unequal intensity in the different phases of the history of the earth, and we have reason to surmise that the percentage of carbon dioxide in the air was considerably greater during periods of strong volcanic activity than it is now, and smaller in quieter periods. Professor Frech, of Breslau, has attempted to demonstrate that this would be in accordance with geological experience, because strongly volcanic periods are distinguished by warm climates, and periods of feeble volcanic intensity by cold climates. The ice age in particular was characterized by a nearly complete cessation of volcanism, and the two periods at the commencement and at the middle of the Tertiary age (Eocene and Miocene) which showed high temperatures were also marked by an extraordinarily developed volcanic activity. This parallelism can be traced back into more remote epochs.

It may possibly be a matter of surprise that the percentage of carbon dioxide in the atmosphere should not constantly be increased, since volcanism is always pouring out more carbon dioxide into our atmosphere. There is, however, one factor which always tends to reduce the carbon dioxide of the air, and that is the weathering of minerals. The rocks which were first formed by the congelation of the volcanic masses (the so-called magma) consist of compounds of silicic acid with alumina, lime, magnesia, some iron and sodium. These rocks were gradually decomposed by the carbonic acid contained in the air and in the water, and it was especially the lime, the magnesia, and the alkalies, and, in some measure also the iron, which formed soluble carbonates. These carbonates were carried by the rivers down into the seas. There lime and magnesia were secreted by the animals and by the algæ, and their carbonic acid became stored up in the sedimentary strata. Högbom estimates that the limestones and dolomites contain at least 25,000 times more carbonic acid than our atmosphere. Chamberlin has arrived at nearly the same figure—from 20,000 to 30,000; he does not allow for the precambrian limestones. These estimates are most likely far too low. All the carbonic acid that is stored up in sedimentary strata must have passed through the atmosphere. Another process which withdraws carbonic acid from the air is the assimilation of plants. Plants absorb carbonic acid under secretion of carbon compounds and under exhalation of oxygen. Like the weathering, the assimilation increases with the percentage of carbonic acid. The Polish botanist E. Godlewski showed as early as 1872 that various plants (he studied *Typha latifolia* and *Glyceria spectabilis* with particular care) absorb from the air an amount of carbonic acid which increases proportionally with the percentage of carbonic acid in the atmosphere up to 1 per cent., and that the assimilation then attains, in the former plant, a maximum at 6 per cent., and in the latter plant at 9 per cent. The assimilation afterwards diminishes if the carbonic acid percentage is further augmented. If, therefore, the percentage of carbon dioxide be doubled, the absorption by the plants would also be doubled. If, at the same time, the temperature rises by 4°, the vitality will increase in the ratio of 1 : 1.5, so that the doubling of the carbon dioxide percentage will lead to an increase in the absorption of carbonic acid by the plant approximately in the ratio of 1 : 3. The same may be assumed to hold for the dependence of the weathering upon the atmospheric percentage of carbonic acid. An increase of the carbon dioxide percentage to double its amount may hence be able to raise the intensity of vegetable life and the intensity of the inorganic chemical reactions threefold.

According to the estimate of the famous chemist Liebig, the quantity of organic matter (freed of water) which is produced by one hectare (2.5 acres) of soil, meadowland, or forest is nearly the same, approximately 2.5 tons per year in central Europe. In many parts of the tropics the growth is much more rapid; in other places, in the deserts and arctic regions, much more feeble. We may be justified in accepting Liebig's figure as an average for the firm land on our earth. Of the organic substances to which we have referred, and which mainly consist of cellulose, carbon makes up 40 per cent. Thus the actual annual carbon production by plants would amount to 13,000 million tons—*i. e.*, not quite fifteen times more than the consumption of coal, and about one-fiftieth of the quantity of the carbon dioxide in the air. If, therefore, all plants were to deposit their carbon in peat-bogs, the air would soon be depleted of its carbon dioxide. But it is only a fraction of one per cent. of the coal which is produced by plants that is stored up for the future in this way. The rest is sent back into the atmosphere by combustion or by decay.

Chamberlin relates that, together with five other American geologists, he attempted to estimate how long a time would be required before the carbon dioxide of the air would be consumed by the weathering of rocks. Their various estimates yielded figures ranging from 5000 to 18,000 years, with a probable average of 10,000 years. The loss of carbonic acid by the formation of peat may be estimated at the same figure. The production of carbonic acid by the combustion of coal would therefore suffice to cover the loss of carbonic acid by weathering and by peat formation seven times over. Those are the two chief factors deciding the consumption of carbonic acid, and we thus recognize that the percentage of carbonic acid in the air must be increasing at a constant rate as long as the consumption of coal, petroleum, etc., is maintained at its present figure, and at a still more rapid rate if this consumption should continue to increase as it does now.

This consideration enables us to picture to ourselves the possibility of the enormous plant-growth which must have characterized certain geological periods of our earth —for instance, the carboniferous period.

This period is known to us from the extraordinarily large number of plants which we find embedded in the clay of the swamps of those days. Those plants were slowly carbonized afterwards, and their carbon is in our age returned to its original place in the household of nature in the shape of carbonic acid. A great portion of the carbonic acid has disappeared from the atmosphere of the earth, and has been stored up as coal, lignite, peat, petroleum, or asphalt in the sedimentary strata. Oxygen was liberated at the same time, and passed into our atmospheric sea. It has been calculated that the amount of oxygen in the air—1216 billion tons—approximately corresponds to the mass of fossil coal which is stored up in the sedimentary strata. The supposition appears natural, therefore, that all the oxygen of the air may have been formed at the expense of the carbonic acid in the air. This view was first advanced by Kœhne, of Brussels, in 1856, and later discussions have strengthened its probability. Part of the oxygen is certainly consumed by weathering processes, and absorbed—*e. g.*, by sulphides and by ferro-salts; without this oxidation the actual quantity of oxygen in the air would be greater. On the other hand, there are in the sedimentary strata many oxidizable compounds—*e. g.*, especially iron sulphides—which have probably been reduced by the interaction of carbon (by or-

CELESTIAL BODIES AS ABODES OF ORGANISMS

ganic compounds). A large number of the substances which consume oxygen during their decomposition and decay have also been produced by the intermediation of the coal which had previously been deposited under liberation of oxygen, so that these substances are, by their oxidation, restored to their original state. We may hence take it as established that the masses of free oxygen in the air and of free carbon in the sedimentary strata approximately correspond to each other, and that probably all the oxygen of the atmosphere owes its existence to plant life. This appears plausible also for another reason. We know for certain that there is some free oxygen in the atmosphere of the sun, and that hydrogen abounds in the sun. The earth's atmosphere may originally have been in the same condition. When the earth cooled gradually, hydrogen and oxygen combined to water, but an excess of hydrogen must have remained. The primeval atmosphere of the earth may also have contained hydrocarbons, as they play an important part in the gases of comets. To these gases there were added carbonic acid and water vapor, coming from the interior of the earth. Thanks to its chemical inertia, the nitrogen of the air may not have undergone much change in the course of the ages. An English chemist, Phipson, claims to have shown that both higher plants (the corn-bind) and lower organisms (various bacteria) can live and develop in an atmosphere devoid of oxygen when it contains carbonic acid and hydrogen. It is also possible that simple forms of vegetable life existed before the air contained any oxygen, and that these plants liberated the oxygen from the carbonic acid exhaled by the craters. This oxygen gradually (possibly under the influence of electric discharges) converted the hydrogen and the hydrocarbons of the air into water and carbonic acid until those elements were consumed. The oxygen remained in the air, whose composition gradually approached more the actual state.[1]

This oxygen is an essential element for the production of animal life. As animal life stands above vegetable life, so animal life could only originate at a later stage than plant life. Plants require, in addition to suitable temperature, only carbonic acid and water, and these gases will probably be found in the atmospheres of all the planets as exhalations of their inner incandescent masses which are slowly cooling. The presence of water vapor has directly been established, by means of the spectroscope, in the atmospheres of other planets—Venus, Jupiter, and Saturn—and indirectly by the observation of a snow-cap on Mars. The spectroscope further gives us indication of the presence of other gases. There is an intense band in the red part of the spectra of Jupiter and

[1] According to the opinion of a colleague of mine, a botanist, the results of the experiments of Phipson must be regarded as very doubtful, and some oxygen would appear to be indispensable for the growth of plants. We have to imagine the development somewhat as follows: As the earth separated from the solar nebula, its temperature was very high at first in its outer portions. At this temperature it was not able to retain the lighter gases, like hydrogen and helium, for a long period; the heavy gases, like nitrogen and oxygen, remained. The original excess of hydrogen and helium disappeared, therefore, before the crust of the earth had been formed, and the atmosphere of the earth immediately after the formation of the crust contained some oxygen, besides much nitrogen, carbonic acid, and water vapor. The main bulk of the actual atmospheric oxygen would therefore have been reduced from carbon dioxide by the intermediation of plants. The view that celestial bodies may lose part of their atmosphere is due to Johnstone Stoney. The atmospheric gases escape the more rapidly the lighter their molecules and the smaller the mass of the celestial bodies. On these lines we explain that the smaller celestial bodies like the moon and Mercury, have lost almost all their atmosphere, while the earth has only lost hydrogen and helium, which again have been retained by the sun.

CELESTIAL BODIES AS ABODES OF ORGANISMS

Saturn, of wave-length 0.000618 mm. Other new constituents of unknown nature have been discerned in the spectra of Uranus and Neptune. On the other hand, there is hardly any, or at any rate only a quite insignificant, atmosphere on the moon and on Mercury. This is easily understood. The temperature on that side of Mercury which is turned away from the sun is near absolute zero. All the gases of the planetary atmosphere would collect and condense there. If, then, Mercury had originally an atmosphere, it must have lost it as it lost its own rotation, compelling it to turn always the same face towards the sun. Similar reasons may account for the absence of a lunar atmosphere. If Venus should likewise always turn the same side towards the sun, as many astronomers assert, Venus should not have any notable atmosphere, nor clouds either. We know, however, that this planet is surrounded by a very marked developed atmosphere.[1]

And that is the strongest objection to the assumption that Venus follows the example of Mercury as regards the rotation about its own axis.

Since, now, warm ages have alternated with glacial periods, even after man appeared on the earth, we have to ask ourselves: Is it probable that we shall in the coming geological ages be visited by a new ice period that will drive us from our temperate countries into the hotter climates of Africa? There does not appear to be much ground for such an apprehension. The enormous combustion of coal by our industrial establishments suffices to increase the percentage of carbon dioxide in the air to a perceptible degree. Volcanism, whose devastations—

---

[1] That results from the very strong refraction which light undergoes in the atmosphere of Venus when this planet is seen in front of the sun's edge during the so-called Venus transits.

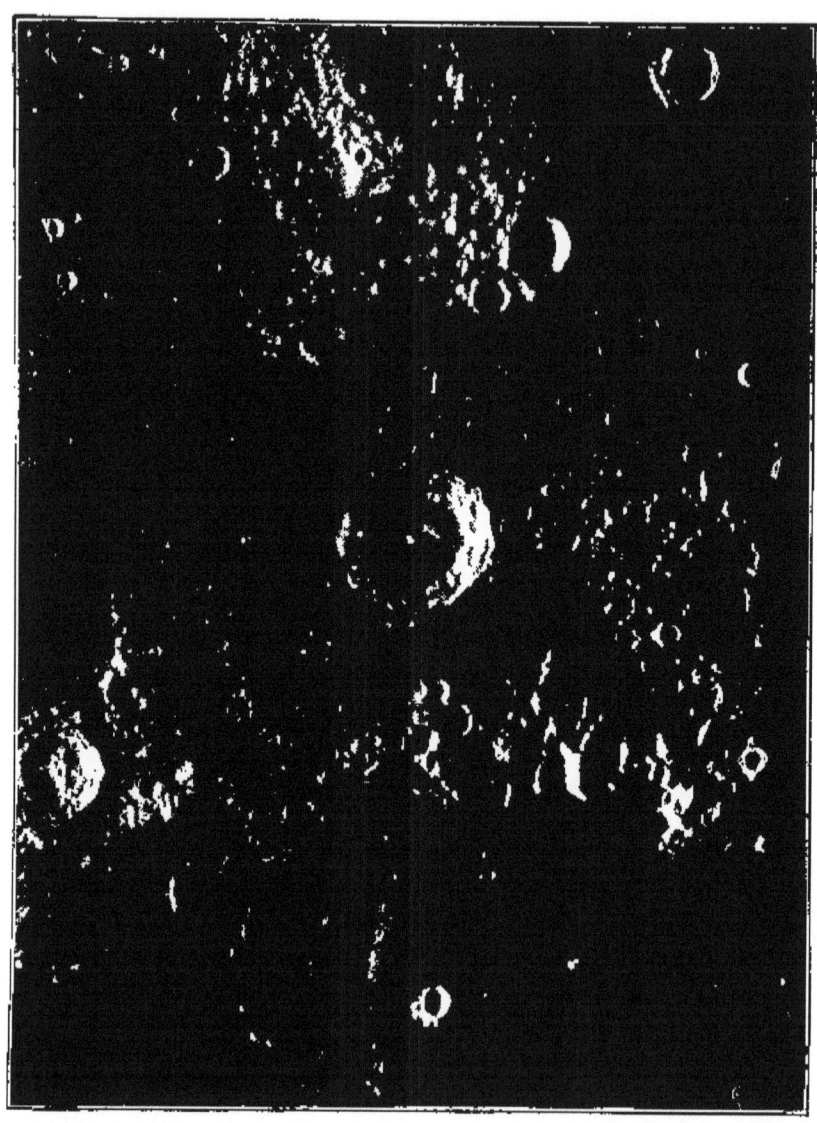

Fig. 17.—Photograph of the surface of the moon, in the vicinity of the crater of Copernicus. Taken at the Yerkes Observatory, Chicago, U. S. A. Scale: Diameter of moon, 0.55 m.=21.7 in. Owing to the absence of an atmosphere and of atmospheric precipitations, the precipitous walls of the crater and other elevations do not indicate any signs of decay

CELESTIAL BODIES AS ABODES OF ORGANISMS

on Krakatoa (1883) and Martinique (1902)—have been terrible in late years, appears to be growing more intense. It is probable, therefore, that the percentage of carbonic acid increases at a rapid rate. Another circumstance points in the same direction; that is, that the sea seems to withdraw carbonic acid from the air. For the carbonic acid percentage above the sea and on islands is on an average 10 per cent. less than the above continents.

If the carbonic acid percentage of the air had kept constant for ages, the percentage of the water would have found time to get into equilibrium with it; but the sea actually absorbs carbonic acid from the air. Thus the sea-water must have been in equilibrium with an atmosphere which contained less carbonic acid than the present atmosphere. Hence the carbonic acid percentage has been increasing of late.

We often hear lamentations that the coal stored up in the earth is wasted by the present generation without any thought of the future, and we are terrified by the awful destruction of life and property which has followed the volcanic eruptions of our days. We may find a kind of consolation in the consideration that here, as in every other case, there is good mixed with the evil. By the influence of the increasing percentage of carbonic acid in the atmosphere, we may hope to enjoy ages with more equable and better climates, especially as regards the colder regions of the earth, ages when the earth will bring forth much more abundant crops than at present, for the benefit of rapidly propagating mankind.

III

RADIATION AND CONSTITUTION OF THE SUN

THE question has often been discussed in past ages, and again in the last century, in how far the position of our earth within the solar system may be regarded as secure. One might apprehend two things. Either the distance of the earth from the sun might increase or decrease, or the rotation of the earth about its axis might be arrested; and either of these possibilities would threaten the continuance of life on the earth. The problem of the stability of the solar system has been investigated by the astronomers, and their patrons have offered high prizes for a solution of the problem. If the solar system consisted merely of the sun and the earth, the earth's existence would be secure for ages; but the other planets exercise a certain, though small, influence upon the movements of the earth. That this influence can only be of slight importance is due to the fact that the total mass of all the planets does not aggregate more than one-seven-hundred-and-fiftieth of the mass of the sun, and, further, to the fact that the planets all move in nearly circular orbits around the centre, the sun, so that they never approach one another closely. The calculations of the astronomers demonstrate that the disturbances of the earth's orbit are merely periodical, representing long cycles of from 50,000 to 2,000,000 years. Thus the whole effect is limited to a slight vacillation of the orbits of the planets about their mean positions.

So far everything is well and good. But our solar system is traversed by other celestial bodies, mostly of unknown, but certainly not of circular orbits—namely, the comets. The fear of a collision with a comet still alarmed the thinkers of the past century. Experience has, however, taught us that collisions between the earth and comets do not lead to any serious consequence. The earth has several times passed through the tails of comets—for instance, in 1819 and 1861—and it was only the calculating astronomer who became aware of the fact. Once on such an occasion we have thought that we observed a glow like that of an aurora in the sky. When the earth was drawing near the denser parts of the comet, particles fell on the earth in the shape of showers of shooting-stars, without doing any appreciable damage. The mass of comets is too small perceptibly to disturb the paths of the planets.

The rotation of the earth about its axis should slowly be diminished by the effects of the tides, since they act like a brake applied to the surface of the earth. This retardation is, however, so unimportant that the astronomers have not been able to establish it in historical times. The slow shrinkage of the earth somewhat counteracts this effect. Laplace believed that we were able to deduce, from an analysis of the observations of solar eclipses in ancient centuries, that the length of the day had not altered by more than 0.01 second since the year 729 B.C.

We know that the sun, accompanied by its planets, is moving in space towards the constellation of Hercules with a velocity of 20 km. (13 miles) per second, which is amazing to our terrestrial conceptions. Possibly the constituents of our solar system might collide with some other unknown celestial body on this journey. But as the celestial bodies are sparsely distributed, we may hope that many billions of years will elapse before such a catastrophe will take place.

In mechanical respects the stability of our system appeared to be well established. Since the modern theory of heat has made its triumphant entry into natural science, however, the aspect of matters has changed. We are convinced that all life and all motion on the earth can be traced back to solar radiation. The tidal motions alone make a rather unimportant exception. We have to ask ourselves: Will not the store of energy in the sun, which goes out, not only to the planets, but to a far greater extent into unknown domains of cold space, come to an end, and will not that be the end of all the joys and sorrows of earthly existence? The position appears desperate when we consider that only one part in 2300 millions of the solar radiation benefits the earth, and perhaps ten times as much the whole system, with all its moons. The solar radiation is so powerful that every gramme of the mass of the sun loses two calories in the course of a year. If, therefore, the specific heat of the sun were the same as that of water, which in this respect surpasses most other substances, the solar temperature would fall by 2° Cent. (3.6° F.) every year. As, now, the temperature of the sun in its outer portion has been estimated at from 6000° to 7000°, the sun should have cooled completely within historical times. And though the interior of the sun most probably has a vastly higher temperature than the outer portions which we can observe, we should, all the same, have to expect that the solar temperature and radiation would noticeably have diminished in historical times. But all the documents from ancient Baby-

RADIATION AND CONSTITUTION OF THE SUN

lon and Egypt seem to point out that the climate at the dawn of historical times was in those countries nearly the same as at present, and that, therefore, the sun shone over the most ancient representatives of culture in the same way as it shines on their descendants now.

The thesis has frequently been advanced, therefore, that the sun has in its heat balance not only an expenditure side, but also an almost equally substantial income side. The German physician R. Mayer, who has the immortal merit of first having given expression to the conception of a relation between heat and mechanical work, directed his attention also to the household of the sun. He suggested that swarms of meteorites, rushing into the sun with an amazing velocity (of over 600 km. per second), would, when stopped in their motion, generate heat at the rate of 45 million calories per gramme of meteorites. In future ages it would be the turn of the planets to sustain for some time longer the spark of life in the sun, by the sacrifice of their own existences. The sun would therefore, like the god Saturn, have to devour its own children in order to continue its existence. Of how little avail that would be we learn from the consideration that the fall of the earth into the sun would not be able to prolong the heat expenditure of the sun by as many as a hundred years. By their rush into the sun, almost uniformly from all sides, the meteorites would, moreover, long since have put a stop to the rotation of the sun about its axis. Further, by virtue of the increasing mass and the hence augmenting attraction of the sun, the length of our year would have had to diminish by about 2.8 seconds per year, which is in absolute contradiction to the observations of the astronomers. According to Mayer's thesis, a corresponding number of meteorites would, finally, also have to tumble upon the surface of the earth, and (according to data which will be furnished in Chapter IV.) they should raise the surface temperature to about 800°. The thesis is therefore misleading.

We must look for another explanation. It occurred to Helmholtz, one of the most eminent investigators in the domain of the mechanical theory of heat, that, instead of the meteorites, parts of the sun itself might fall towards its centre, or, in other words, that the sun was shrinking. Owing to the high gravitation of the sun (27.4 times greater than on the surface of the earth), the shrinkage would liberate a great amount of heat. Helmholtz calculated that, in order to cover the heat expenditure of the sun, a shrinkage of its diameter by 60 m. annually would be required. If the sun's diameter should only be diminished by one-hundredth of one per cent.—a change which we should not be able to establish—the heat loss would be covered for more than 2000 years. That seems at first satisfactory. But if we proceed with our estimate, we find that if the sun went on losing as much heat as at present for seventeen million years it would have to contract within this period to a quarter of its present volume, and would therefore acquire a density like that of the earth. Long before that, however, the radiation from the sun would have been decreased so powerfully that the temperature on the earth's surface would no longer rise above freezing-point. Helmholtz, on this argument, limited the further existence of the earth to about six million years. That is less satisfactory. But we know nothing of the future and must be content with possibilities. Not so, however, if we calculate back with the aid of Helmholtz's theory. According to this theory, and according to Helmholtz's own data, a state like the present cannot have existed for more than ten million years.

RADIATION AND CONSTITUTION OF THE SUN

Since, now, geologists have come to the conclusion that the petrefactions which we find in the fossil-bearing strata of the earth have needed at least a hundred million years for their formation, and more probably a thousand million years, and since, moreover, the still more ancient formations—the so-called precambrian strata—have been deposited in equally long or still longer periods, we see that the theory of Helmholtz is unsatisfactory.

A somewhat peculiar way out of the dilemma has been suggested by a few scientists. We know that one gramme of the wonderful element radium emits about 120 calories per hour, or in the course of a year, in round numbers, a million calories. This radiation seems to continue unimpaired for years. If we now assume that each kilogramme of the mass of the sun contains only two milligrammes of radium, that amount would be sufficient to balance the heat expenditure of the sun for all future ages. Without some further auxiliary hypothesis, we can, however, not listen to this suggestion. It presupposes that heat is created out of nothing. Some scientists, indeed, believe that radium may absorb a radiation, coming from space, in some unknown manner and convert it into heat. Before we enter seriously into a discussion of this explanation we shall have to answer the questions where that radiation comes from and where it takes its store of energy.

We must, therefore, again search for another source of heat energy for the sun. Before we can hope to find it, we had better study the sun itself a little.

All scientists are agreed that the sun is of the same constitution as the thousands of luminous stars which we see in the sky. According to the color of the light which they emit, stars are classified as white, yellow, and red stars. The differences in their light become much more distinct when we examine them spectroscopically. In the white stars the helium and hydrogen lines predominate decidedly; the helium stars contain, in addition, oxygen. Metals are comparatively little represented; but they play a main part in the spectra of the yellow stars, in which, further, some bands become visible. In the spectra of the red stars we notice many bands which indicate that chemical compounds are present in the outer portions. Everybody knows that the platinum wire or the filament of an incandescent lamp which has been heated to incandescence by the electric current first shines reddish, then yellow when the current is increased, and finally more and more white. At the same time the temperature rises. We can estimate the temperature from the brightness of the glow. If we know the wave-length of the radiations of that color which emits the greatest amount of heat in the spectrum (it should be a normal spectrum), it is easy to calculate the temperature of the star from Wien's law of displacements. We need only divide 2.89 by the respective wave-length expressed in mm. to find the absolute temperature of the star; by deducting 273 from the result, we obtain the temperature in degrees Cent. on the ordinary scale. For the sun the maximum of heat radiation lies near wave-length 0.00055 (in the greenish-yellow light), and therefore the absolute temperature of the radiating disk of the sun, the so-called photosphere, should be $5255°$ absolute, or nearly $5000°$ Cent. But our atmosphere weakens the sunlight, and it also causes a displacement of the maximum radiation in the spectrum. The same applies to the sun's own atmosphere, so that we have to adopt a higher estimate than $5000°$ Cent. By means of Stefan's law of radiation, the solar temperature has been estimated at about $6200°$, which would corre-

RADIATION AND CONSTITUTION OF THE SUN

spond to a wave-length of about 0.00045 mm. This correction is therefore significant. About half of it has to be ascribed to the influence of the solar atmosphere, the other half to the terrestrial atmosphere. A Hungarian astronomer, Harkányi, has determined in the same way the temperature of several white stars (Vega and Sirius), and found it to be about 1000° higher than that of the sun, while the red star Betelgeuse, the most prominent star in Orion, would have a temperature by 2500° lower than that of the sun.

It must expressly be stated that in making these estimates we understand by the temperature of the star in this case the temperature of a radiating body which emits the same light as that which reaches us from the star. But the stellar light undergoes important changes on its way to us. We learn from observing new stars that a star may be surrounded by a cloud of cosmical dust which sifts the blue rays out and permits the red ones to pass. The star then shines with a less brilliantly white light than in the absence of the cloud. The consequence is that we estimate the temperature lower than it really is. In the red stars bands have been noticed, indicating, as we have already said, the presence of chemical compounds. The most interesting of these are the compounds of cyanogen and of carbon, probably with hydrogen, which appear to resemble those observed by Swan in the spectrum of gas flames and which were named after him. It was formerly thought that the presence of these compounds implied lower temperature. But we shall see that this conclusion is not firmly established. Hale has found during eclipses of the sun that exactly the same compounds occur immediately above the luminous clouds of the sun. They are probably more numerous below the clouds, where the temperature is no doubt higher, than above them.

However that may be, we have reason to assume that the now yellow sun was once a white star like the brilliant Sirius, that it has slowly cooled down to its present appearance, and that it will some day shine with the reddish light of Betelgeuse. The sun will then only radiate a seventh of the heat which it emits now, and it is very likely that the earth will have been transformed into a glacial desert long before that time.

It has already been pointed out that the atmospheres of both the sun and of the earth produce a strong absorption of the solar rays, and especially of the blue and white rays. It is for this reason that the light of the sun appears more red in the evening than at noon, because in the former case it has to pass through a thicker layer of air, which absorbs the blue rays. For the same reason the limb of the sun appears more red in spectroscopic examinations than the centre of the sun. This weakening of the sun's light is due to the fine dust pervading the atmospheres of the earth and the sun. When the products of strong volcanic eruptions, like the eruptions of Krakatoa in 1883 and of Mont Pelée in 1902, filled the atmosphere with a fine volcanic dust, the sun appeared distinctly red when standing low in the horizon. It was this dust that caused the red glow.

When we examine an image of the sun which has been thrown on a screen by the aid of a lens or a system of lenses, we notice on the sun's disk a mottling of characteristic darker spots. These spots struck the attention of Galileo, and they were discovered almost simultaneously by him, by Fabricius, and by Scheiner (1610-1611). These spots have since been the most diligently studied features of the sun. We carefully determine their number and sizes, and combine these two data to make the so-called sun-spot numbers. These numbers change

RADIATION AND CONSTITUTION OF THE SUN

from year to year in a rather irregular way, the period amounting on an average to 11.1 years. The spots appear in two belts on the sun, and they glide over the disk in the course of thirteen or fourteen days. Sometimes they reappear after another thirteen or fourteen days. It is therefore believed that they lie comparatively quiet on the surface of the sun, and that the sun rotates about its own axis in about twenty-seven days, so that after that period the same points are again opposite the earth. This is the so-called synodical period. The great interest which attaches to the study of these features lies in the fact that simultaneously with these spots several other phenomena seem to vary which attain their maxima at the same time. Such are, in the first instance, the polar lights and the magnetic variations, and, to a lesser degree, the cirrus clouds and temperature changes, as well as several other meteorological phenomena (compare Chapter V.).

About the sun-spots we notice the so-called faculæ—portions which are much brighter than their surroundings. When we carefully examine a strongly magnified image of the sun, we find that it has a granulated appearance (Fig. 18). Langley compares the disk to a grayish-white cloth almost hidden by flakes of snow. The less bright portions are designated "pores," the brighter portions "granules." It is generally assumed that the granules correspond to clouds which rise like the clouds of our atmosphere on the top of ascending convection currents. But while the terrestrial clouds are formed of drops of rain or of crystals of ice, the granules consist probably of soot—that is to say, condensed carbon—and of drops of metals, iron, and others. The smallest granule which we are able to discern has a diameter of about 200 km. (130 miles).

The faculæ are formed by very large accumulations of

Fig. 18.—Sun-spot group and granulation of the sun. (Photographed at the Meudon Observatory, near Paris, April 1, 1884)

clouds which are carried up by strong ascending currents and spread over large areas, as in our cyclones. The spots correspond to descending masses of gas with rising temperatures, which are therefore "dry" and do not carry any clouds, as in terrestrial anticyclones. Through these holes in the walls of solar clouds we peep a little farther into the gigantic masses of gas, and we obtain an idea of the state of affairs in the deeper strata of the sun. The depth of the wall of cloud is, of course, not large compared to the radius of the sun.

The study of the spectra affords us the best insight into the nature of the different parts of the sun. The spectra teach us not only the constituents of these parts, but also the velocities with which they move. We have learned in this way that, lying above the luminous clouds of the sun which are radiating to us, there are great masses of gas containing most of our terrestrial elements. We distinguish particularly in them iron, magnesium, calcium, sodium, helium, and hydrogen. The two last-mentioned constituents, being the least dense, are found particularly in the outermost strata of the atmosphere. The solar atmosphere becomes visible when, during an eclipse of the sun, the disk of the moon has proceeded so far as to cover the intensely luminous clouds in the so-called photosphere. Owing to its strong percentage of

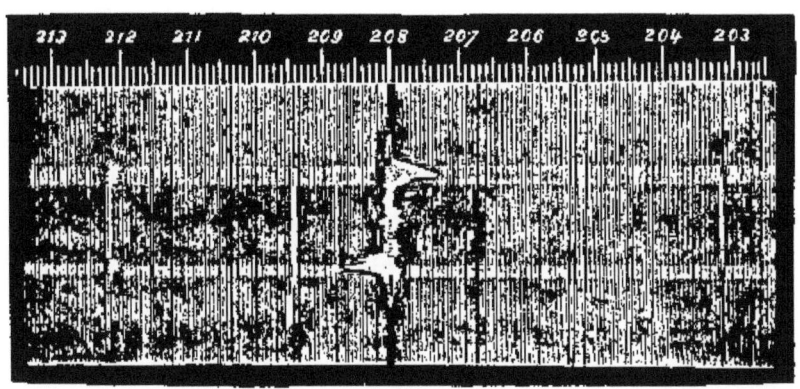

Fig. 19.—Part of the solar spectrum of January 3, 1872. After Langley. The bright horizontal bands are due to prominences. In the middle (at 208) the hydrogen line F, strongly distorted by violent agitation

hydrogen, the gaseous atmosphere generally shines in the purple hue which is characteristic of this element. This stratum of gas is also called the chromosphere (from the Greek word χρῶμα, meaning color). Its thickness is estimated at from 7000 to 9000 km. (5000 to 6000 miles).

Fig. 20. — Metallic prominences in vortex motion
The white spot marks the size of the earth

Fig. 21.—Fountain-like metallic prominences

From it rise rays of fire over the surrounding surface like blades of grass on meadows, to which their appearance has been likened.

When these flames rise still higher, to about 15,000 km. (9300 miles) or more, they are called protuberances or prominences. Their number as well as their altitude grow with the number of sun-spots. They are distinguished as metallic and as quiet prominences. The former are characterized by particularly violent motion, as will become apparent from Figs. 20 and 21, and they contain large amounts of metallic vapor. They appear only within the belt of sun-spots which are most pronounced at a distance of about 20° from the solar equator. Their movements are so violent that they often traverse several hundreds of kilometres in a second. The Hungarian Fényi observed, indeed, on July 15, 1895, a prominence whose greatest velocity in the line of sight, measured spectroscopically, amounted to 862 km. (536 miles), and whose maximum velocity at right angles to this direction was 840 km. per second. These colossal velocities distinguish the highest parts, while the lower portions, which are the most dense and which contain most metallic vapor, are less mobile, as might be expected. Their altitude above the sun's surface may reach exceedingly high figures, and this applies also to the quiet prominences. The above-mentioned prominence of July 15, 1895, reached a height of 500,000 km., and Langley observed, on October 7, 1880, one at an altitude of 560,000 km., whose tip, therefore, nearly attained an elevation equal to that of a radius of the sun, 690,000 km. above the limb of the sun's photosphere. The mean altitude of these prominences is 40,000 km. After their discovery by Lector Vassenius, of Götheborg, in 1733, they could only be studied during total solar eclipses, until

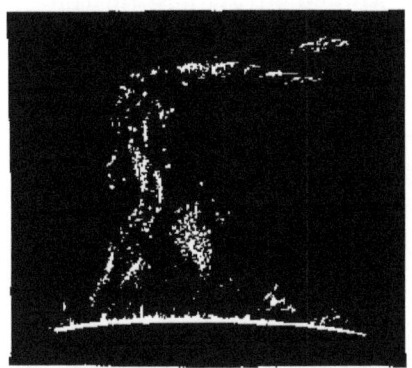
Fig. 22.—Quiet prominences of smoke-column type

Fig. 23.—Quiet prominences, shape of a tree. The white spot indicates the size of the earth

Lockyer and Janssen taught us, in the year 1868, how to observe them in full sunlight by means of the spectroscope.

The quiet prominences consist almost exclusively of

hydrogen and helium; sometimes they contain also traces of metallic gases. They resemble clouds floating quietly in the solar atmosphere, or masses of smoke com-

Fig. 24.—Diagram illustrating the differences in the spectra of sun-spots and of the photosphere. Some lines in the spot spectrum are stronger, others fainter, than in the photosphere spectrum. In the central portion, two reversals; to the right, two bands. After Mitchell

ing from a chimney. They may appear anywhere on the sun, and their stability is so great that they have sometimes been watched during a complete solar rotation (for

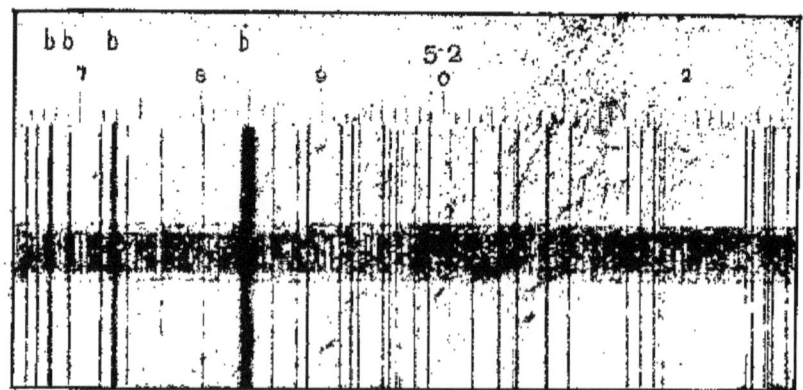

Fig. 25.—Spectrum of a sun-spot, the central band between the two portions of the photosphere spectrum. The spot spectrum is bordered with the half-shadows of the edge of the spot. After Mitchell

about forty days); this is possible only when they occur in the neighborhood of the poles, where they always remain visible outside the sun's limb. Figs. 22 and 23 show several such prominences according to Young.

Sometimes the matter of the prominences seems to fall

Fig. 26.—The great sun-spot of October 9, 1903. Taken with the photo-heliograph of Greenwich in the usual manner. The spot is shown at mean level of the calcium faculæ. The two following photographs show a lower-level and a higher-level section through the calcium faculæ

back upon the surface of the sun between the smaller flames of fire which we have likened to blades of grass (Fig. 21). In most cases, however, the prominences appear slowly to dissolve. When their brilliant glow fades owing to their intense radiation, they can no longer be observed. The quiet prominences, which seem to float at heights of about 50,000 km. and at still greater heights, must there be almost in a vacuum. Their particles cannot be supported by any surrounding gases, after the manner of the drops of water in terrestrial clouds. In order that they may remain floating they must be pushed away from the sun by a peculiar force— the radiation pressure (see Chapter IV.).

The faculæ can be studied in the same way as the prominences, and of late Deslandres and Hale have used for this purpose a special instrument, the heliograph (compare Figs. 26 to 29). When the faculæ approach the limb of the sun they appear particularly brilliant by comparison with their surroundings. That seems to indicate that they are lying at a great altitude, and that

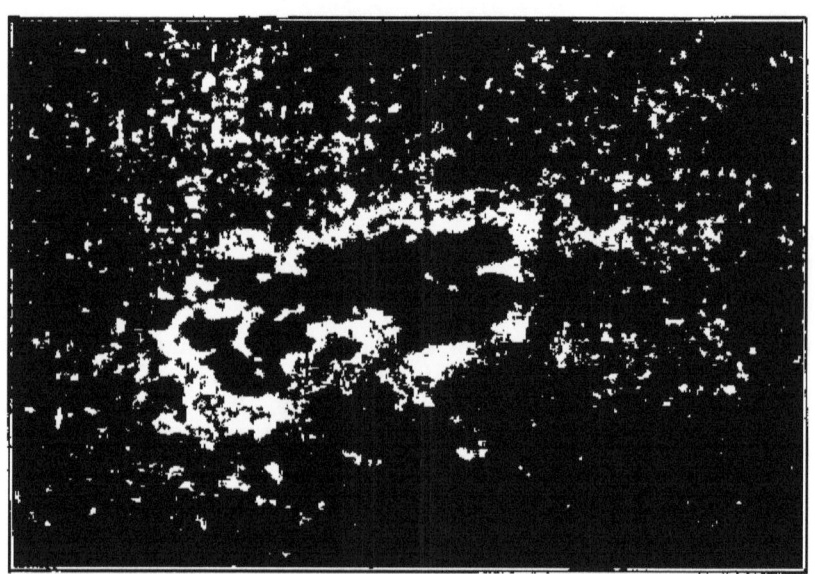

Fig. 27.—The great sun-spot of October 9, 1903. Photograph of the low-level calcium faculæ with the aid of the light of the calcium line H. The spot is not obscured by the faculæ -at least, not so much as in the following illustrations

RADIATION AND CONSTITUTION OF THE SUN

their light is hence not weakened by the superposed hazy stratum. When they reach the sun's limb they appear to us like raised portions of the photosphere. The clouds

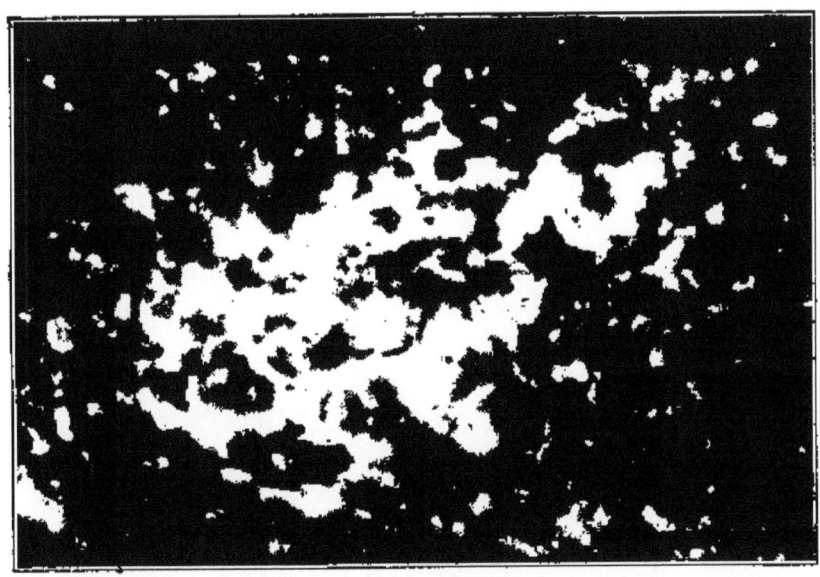

Fig. 28.—The great sun-spot of October 9, 1903. Photograph of the higher-level calcium faculæ, taken with the light of the central portion of the line H (calcium). The higher-level faculæ hide the spot, indicating that the faculæ spread considerably during their ascent

which form these faculæ are carried upward by powerful ascending streams of gas whose expansion is due to the diminution of the gaseous pressure.

Sun-spots display many peculiarities in their spectra (Figs. 24 and 25). Very prominent is always the helium line; prominent likewise the dark sodium lines, which are markedly widened and which show in their middle portions a bright line—the so-called reversal of lines (Fig. 24). This occurrence indicates that the metal is lying in a deeper stratum. In the red portion of the spectrum we find bands, just as in the spectra of the red stars. These bands, which appear to be resolved into crowds of lines by the aid of powerful instruments, indicate the presence of chemical compounds. Since the spot is comparatively of feeble intensity, its spectrum appears superposed like a less bright ribbon upon the background of the spectrum of the more luminous photosphere. The violet end of the sun-spot spectrum is particularly weakened. Although the spot has the appearance of a pit in the

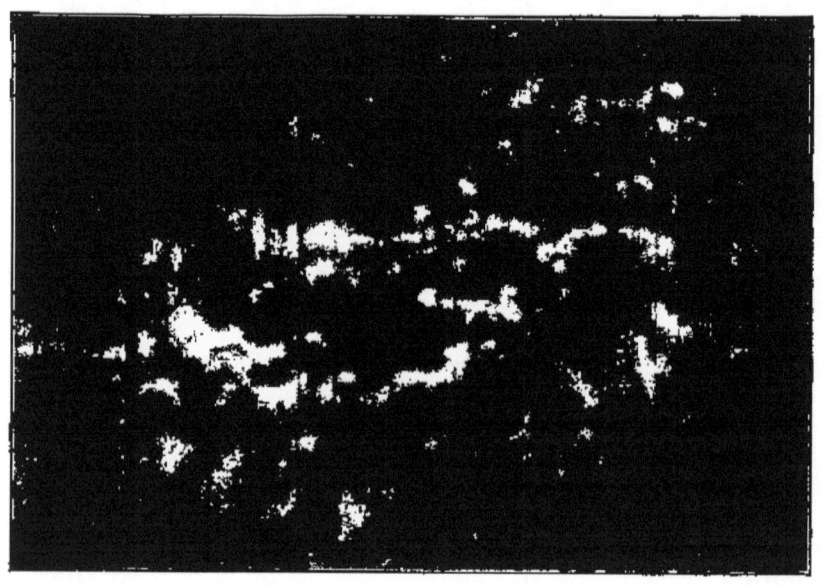

Fig. 29.—The great sun-spot of October 9, 1903. Photograph of the hydrogen faculæ, taken with the light of the spectral line F (hydrogen). Only the darkest portions of the spot are visible. The other portions are obscured by masses of the hydrogen, which were evidently in a restless state

photosphere, and when on the sun's limb makes it look as if a piece had been cut out of the edge, it yet does not appear darker than the sun's edge. That points

Fig. 30.—Photograph of the solar corona of 1900. (After Langley and Abbot.) Illustrating the appearance of the corona in years of minimum sun-spot frequency

to the conclusion that the light emitted by the spot emanates chiefly from its upper, cold portions.

The light coming from the deeper portions is distinctly absorbed to a large degree by the higher-lying strata. The sun-spots also appear to become narrower in their lower parts, owing to the compression of the gases at greater depths, and one may regard their funnel-shaped cloud-walls as "half-shadows," which appear darker than the surroundings, but brighter than the so-called core of the spot. The weakening of the violet end of the spectrum is probably due to the presence of fine particles of dust in the solar gases, just as they cause the corresponding weakening of the violet end of the spectrum of the sun's limb. The bands in the red parts of the sun-spot spectrum may originate from the deeper portions of the spot, because all the higher parts of the solar

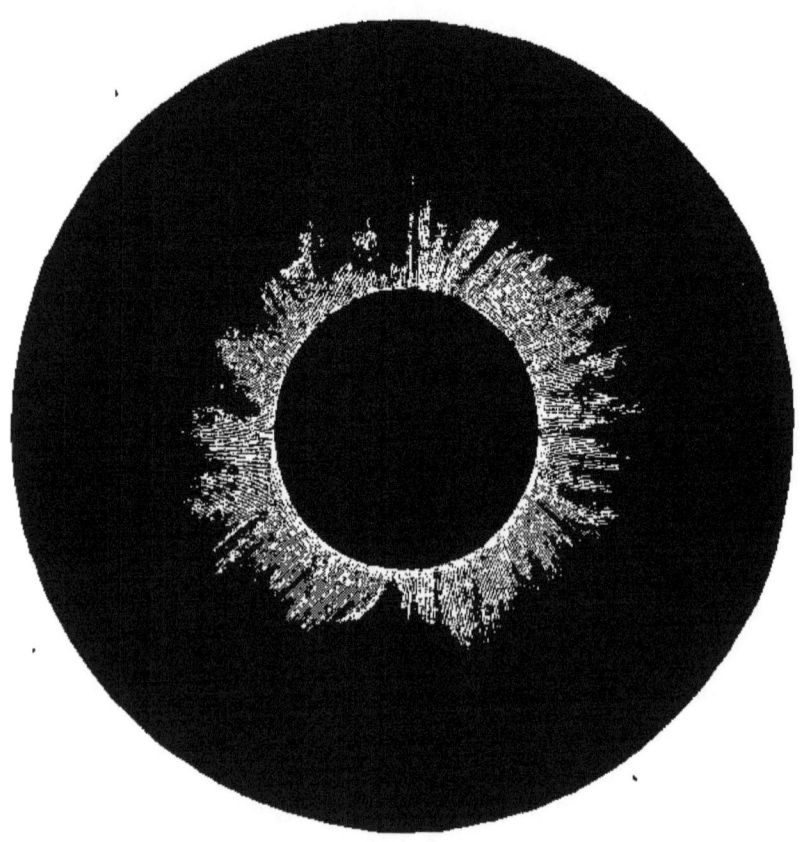

Fig. 31.—Photograph of the solar corona of 1870. (After Davis.)
The year 1870 was one of maximum sun-spot frequency

atmosphere yield simple, sharp lines. The bands suggest that chemical compounds can exist at the higher pressure of the inner portions of the sun, and that these compounds are decomposed in the outer parts of the sun, to give the line spectra of chemical elements.

The enigmatical corona lies farther out in the atmosphere of the sun. It consists of streamers which may extend beyond the disk of the sun to the length of several solar diameters. The corona can only be observed at total eclipses of the sun. Figs. 30 to 32 illustrate the appearance of this very peculiar phenomenon.

When the number of sun-spots is small, the corona streamers extend like huge brooms from the equatorial

Fig. 32.—Photograph of the solar corona of 1898. (After Maunder.) 1898 was a year of average solar activity

parts, and the feebler rays of the corona near the solar poles are then bent downward to the equator, just like the lines of force about the poles of a magnet (Fig. 30).

We suppose, for this reason, that the sun acts like a strong magnet, whose poles are situated near the geographical poles of the sun. In years which are richer in sun-spots the distribution of the streamers of the corona is more uniform. At moderate sun-spot frequency, large numbers of rays seem to emanate from the neighborhood of the maximum belt of sun-spots, so that the corona often assumes a quadrangular shape (compare Fig. 32).

These remarks hold for the "outer corona," while the inner portion, the so-called "inner corona," shines in a more uniform light. The spectroscopic examination demonstrates that the light consists mainly of hydrogen gas and of an unknown gas designated coronium, which particularly seems to occur in the higher parts of the inner corona. The outer streamers of the corona, on the contrary, yield a continuous spectrum which shows that the light is radiated by solid or liquid particles. In the spectrum of the coronal rays at an extreme distance from the disk, astronomers have sometimes fancied that they discerned dark lines on a bright ground, just as in the spectrum of the photosphere. It has been assumed that this light is reflected sunlight, originating from small solid or liquid particles of the outer corona. It must be reflected, because it is partly polarized. The radiating disposition of the outer corona indicates the action of a force, the radiation pressure, which drives the smaller particles away from the centre of the sun.

As regards the temperature of the sun, we have already seen that the two methods applied for its determination have yielded somewhat unequal results. From the intensity of the radiation, Christiansen, and afterwards Warburg, calculated a temperature of about 6000° Cent. Wilson and Gray found for the centre of the sun 6200°, which they afterwards corrected into 8000°. Owing to the absorption of light by the terrestrial and the solar atmospheres, we always find too low values. That applies, to a still greater extent, to any estimate based upon the determination of that wave-length for which the heat emission from the solar spectrum is maximum. Le Chatelier compared the intensity of sunlight filtered

RADIATION AND CONSTITUTION OF THE SUN

through red glass with the intensities of light from several terrestrial sources of fairly well-known temperatures treated in the same way. These estimates yielded to him a solar temperature of 7600° Cent. Most scientists reckon with an absolute temperature of 6500°, corresponding to about 6200° Celsius. That is what is known as the "effective temperature" of the sun. If the solar rays were not partially absorbed, this temperature would correspond to that of the clouds of the photosphere. Since red light is little absorbed comparatively, Le Chatelier's value of 7600°, and the almost equal value of Wilson and Gray of 8000°, should approximately represent the average temperature of the outer portions of the clouds of the photosphere. The higher temperature of the faculæ is evident from their greater light intensity, which, however, may partly be due to their greater height. Carrington and Hodgson saw, on September 1, 1859, two faculæ break out from the edge of a sun-spot. Their splendor was five or six times greater than that of the surrounding parts of the photosphere. That would correspond to a temperature of about 10,000 or 12,000° Cent. The deeper parts of the sun which broke out on these occasions evidently have a higher temperature, and this is not unnatural, since the sun is losing heat by radiation from its outer portions.

We know that the temperature of our atmosphere decreases with greater heights. The movements of the air are concerned in this change. A sinking mass of air is compressed by the increased pressure to which it is being exposed, and its temperature rises, therefore, just as the temperature rises in a pneumatic gas-lighter when the piston is pressed down. If the air were dry and in strong vertical motion, its temperature would change by 10° Cent. (18° F.) per km. If it stood still, it would assume an almost uniform temperature; that is to say, there would be no lowering of the temperature as we proceed upward. The actual value lies between the two extremes. As the gravitation in the photosphere of the sun is 27.4 times greater than on the surface of the earth, we can deduce that, if the air on the sun were as dense as on the earth, the temperature on the sun would vary 27.4 times as much as on the earth with the increasing height—that is to say, by 270 degrees per kilometre, provided its atmosphere were in violent agitation. Now, the outer portions of the solar atmosphere are, indeed, in violent motion, so that this latter assumption seems to be justified. But this part consists essentially of hydrogen, which is 29 times lighter than the air. We must, therefore, reduce the value at which we arrive to one-twenty-ninth. As a result, the final temperature gradient per kilometre would only be 9° Cent. (16.2° F.). But the radiation is extremely powerful on the sun, and it tends to equalize the conditions. Nine degrees per kilometre is therefore, without doubt, too high a value. Further, in the interior of the sun the gases are much heavier. At a small depth, however, they will be so strongly compressed by the upper strata that their further compressibility will be limited, and the calculation which we have just made loses its validity. Yet, in any case, the temperature of the sun must increase as we penetrate nearer to its centre. If we accept a temperature gradient per kilometre of the value above indicated, 9°—it is three times greater in the solid earth-crust—we should obtain for the centre of the sun a temperature of more than six million degrees.

All substances melt and evaporate as their temperature is raised. If the temperature exceeds a certain limit, the "critical temperature," the substance can no longer be condensed to a liquid, however high the pressure may be pushed, and the substance will only exist as a gas. If we start from —273° as absolute zero, this critical temperature is nearly one and a half times as high as the ebullition temperature of the substance under atmospheric pressure. So far as our experience goes, it does not appear probable that the critical temperature of any substance could be higher than 10,000° or 12,000° Cent., the highest values which we have calculated for the temperature of the faculæ. The inner portions of the sun must hence be gaseous, and the whole sun be a strongly compressed mass of gas of extremely high temperature, which, owing to the high pressure, is at a density 1.4 times as great as that of water, and which in many respects, therefore, will resemble a liquid. It must, for instance, be extremely viscid, and that accounts for the relatively great stability of the sun-spots (one sun-spot held out for a year and a half in 1840 and 1841). The sun would thus have to be regarded as a sphere of gas, in the outer portions of which a certain amount of condensations of cloud character have taken place, owing to radiation and to the outward movements of the gaseous masses. The pressure in the photosphere — that is, in those parts in which these clouds are floating—has been averaged at five or six atmospheres, a figure which, considering the very high gravitation, would suggest a layer of superposed gas above it corresponding to not more than a fifth of our terrestrial atmosphere. At an approximately corresponding height, 11,500 m. (38,000 ft.), there are floating in the terrestrial atmosphere the highest cirrus clouds, to which the clouds of the photosphere may in many respects be compared.

We turn back to the unanswered question whence the sun takes the compensation for the heat which it constantly radiates into space. The most powerful source of heat known to us is that of chemical reactions. The most familiar reaction of daily life is the combustion of coal. By burning one gramme of carbon we obtain 8000 calories. If the sun consisted of pure carbon, its energy would not hold out more than 4000 years. It is not to be wondered at, therefore, that most scientists soon abandoned the hope of solving the problem in this way. The French astronomer Faye attempted to explain the replenishment of the losses of heat by radiation from the sun by arguments in which he resorted to the heat of a combination of the constituents of the sun. He said: "So high a temperature must prevail in the interior of the sun that everything there will be decomposed into its elementary constituents. When the atoms afterwards penetrate into the outer layers, they are again united, and they liberate heat." Faye thus imagined that new masses of elements would constantly rise from the interior of the sun and would be reunited in chemical combination on the surface. But if new masses are to penetrate upward to the surface, those which were at first above must go back to the centre of the sun, in order to be re-decomposed by the great heat there; and this re-decomposition would consume just as much heat as was gained by the rising of the same masses to the surface. This convection can therefore only help to transport the store of heat from the interior to the surface. The total amount of heat stored in the sun would in this way, supposing the mean temperature to be six million degrees, be able to cover the heat expenditure for about three million years.

We have, moreover, seen that the highest strata of the sun are distinguished by line spectra, suggestive of simple chemical compounds, while at greater depth in the sun-spots chemical combinations occur which are characterized by band spectra. It is quite incorrect to assert that high temperatures must necessarily decompose all chemical compounds into their elements. The mechanical theory of heat teaches us only that at rising temperatures products are formed whose formation goes hand in hand with an absorption of heat. Thus, at a high temperature, ozone is formed from oxygen, although ozone is more complex in composition than oxygen, and by this reaction 750 calories are consumed when one gramme of oxygen is transformed into one gramme of ozone. We likewise know that in the electric arc, at a temperature of about 3000°, a compound is formed under consumption of heat by the oxygen and nitrogen of the atmosphere. A new method for the technical preparation of nitric acid from the nitrogen of the air is based upon this reaction. Again, the well-known compounds benzene and acetylene are formed from their elements, carbon and hydrogen, under absorption of heat. All these bodies can only be synthetized from their elementary constituents at high temperatures. We further know from experience that the higher the temperature at which a reaction takes place, the greater, in general, the amount of heat which it absorbs.

A similar law applies to the influence of pressure. When the pressure is increased, such processes will be favored as will yield products of a smaller volume. If we imagine that a mass of gas rushes down from a higher stratum of the sun into the depths of the sun's interior, as gases do in sun-spots, complex compounds will be produced by virtue of the increased pressure. This pressure must increase at an immense rate towards the interior of the sun, by about 3500 atmospheres per kilometre. The gases which dissociate into atoms at the lower pressures and the higher temperatures of the extreme solar strata above the photosphere clouds enter into chemical combination in the depths of the spots, as we learn from spectroscopic examination. Owing to their high temperatures, these compounds absorb enormous quantities of heat in their building up, and these quantities of heat are to those which are concerned in the chemical processes of the earth in the same ratio as the temperature of the sun is to that at which the chemical reactions are proceeding on the earth. As these gases penetrate farther into the sun, temperature and pressure are still more and more increased, and there will result products more and more abounding in energy and concentration. We may, therefore, imagine the interior of the sun charged with compounds which, brought to the surface of the sun, would dissociate under an enormous evolution of heat and an enormous increase of volume. These compounds have to be regarded as the most powerful blasting agents, by comparison with which dynamite and gun-cotton would appear like toys. In confirmation of this view, we observe that gases when penetrating into the photosphere clouds are able to eject prominences at a stupendous velocity, attaining several hundred kilometres per second. This velocity surpasses that of the swiftest rifle-bullet about a thousandfold. We may hence ascribe to the explosives which are confined in the interior of the sun energies which must be a million times greater than the energy of our blasting agents. (For the energy increases with the square of the velocity.) And yet these solar blasting agents have already given up a large part of their energy during their passage from the sun's interior. It thus becomes conceivable that the solar energy—instead of holding out for 4000 years, as it would if it depended upon the combustion of a solar

RADIATION AND CONSTITUTION OF THE SUN

sphere made out of carbon—will last for something like four thousand million years. Perhaps we may further extend this period to several billions.

That there are such energetic compounds we have learned from the discovery of the heat evolution of radium. According to Rutherford, radium is decomposed by one-half in the space of about ~~1800~~ years. In this decomposition a quantity of about a million calories is evolved per gramme and per year, and we thus find that the decomposition of radium into its final products is accompanied by a heat evolution of about two thousand millions of calories per gramme—about a quarter of a million times more heat than the combustion of one gramme of carbon would yield.

In chemical respects as well, then, the earth is a dwarf compared to the sun, and we have every reason to presume that the chemical energy of the sun will be sufficient to sustain the solar heat during many thousand millions and possibly billions of years to come.

IV

THE RADIATION PRESSURE

NEXT to simple measuring and simple calculations, astronomy appears to be the most ancient science. Yet, though man has worshipped the sun from the most remote ages, it was not fully comprehended before the middle of the past century that the sun is the source of all life and of all motion. Part of the veneration for the sun was transferred to the moon, with its mild light, and to the smaller celestial lights. It did not escape notice that their positions in the sky were always changing simultaneously with the annual variations in the weather, and all human undertakings depended upon the weather and the seasons. The moon and the stars were worshipped—we know now, without any justification whatever—as ruling over the weather, and consequently over man's fate.[1] Before anything was undertaken people attempted first to assure themselves of the favorable aspect of the constellations, and since the most remote ages astrologers have exercised a vast influence over the ignorant and superstitious multitude.

In spite of the vehement enunciation of Giordano

---

[1] The moon strongly, and more than any other agent, influences the tides. Apart from this effect the position of the moon has only a feeble influence upon the air pressure and upon atmospheric electricity and terrestrial magnetism. The influence of the stars is imperceptible.

Bruno (1548–1600), this superstition was still deeply rooted when Newton succeeded in proving, in 1686, that the movements of the so-called wandering stars, or planets, and of their moons could be calculated with the aid of one very simple law: that all these celestial bodies are attracted by the sun or by their respective central bodies with a force which is proportional to their own mass and to the mass of the central body and inversely proportional to the square of their distance from that central body. Newton's contemporary, Halley, applied the law of gravitation also to the mysterious comets, and calculating astronomy has since been based upon this, its firmest law, to which there has not been found any exception. The world was thus at once rid of the paralyzing superstition which exacted belief in a mysterious ruling of the stars. The contemporaries of Newton, as well as their descendants, have rightly valued this discovery more highly than any other scientific triumph of this hero's. According to Newton's law, all material bodies would tend to become more and more concentrated and united, and the development of the universe would result in the sucking up of the smaller celestial bodies—the meteorites, for instance—by the larger bodies.

It must, however, be remarked that Newton's great precursor, Kepler, observed in 1618 that the matter of the comets is repelled by the sun. Like Newton, he believed in the corpuscular theory of light. The sun and all other luminous bodies radiated light, they thought, because they ejected minute corpuscles of light matter in all directions. If, now, these small corpuscles hit against the dust particles in the comets' tails, the dust particles would be carried away with them, and their repulsion by the sun would become intelligible. It is characteristic that Newton would not admit this explanation of Kepler's, although he shared Kepler's opinion on the nature of light. According to Newton, the deviation of the tails of comets from his law of general attraction was only apparent. The tails of comets, he argued, behaved like the columns of smoke rising from a chimney, which, although the gases of combustion are attracted by the earth, yet ascend because they are lighter than the surrounding air. This view, which has been characterized by Newcomb as no longer to be seriously taken into consideration, demonstrates the strong tendency of Newton to explain everything with the aid of his law.

The astronomers followed faithfully in the footsteps of their inimitable master, Newton, and they brushed aside every phenomenon which would not fit into his system. An exception was made by the famous Euler, who, in 1746, expressed the opinion that the waves of light exerted a pressure upon the body upon which they fell. This opinion, however, could not prevail against the criticisms with which others, and especially De Mairan, assailed it. That Euler was right, however, was proved by Maxwell's great theoretical treatise on the nature of electricity (1873). He showed that rays of heat—and the same applies, as Bartoli established in 1876, to radiations of any kind—must exercise a pressure just as great as the amount of energy contained in a unit volume, by virtue of their radiation. Maxwell calculated the magnitude of this pressure, and he found it so small that it could hardly have been demonstrated with the experimental means then at our disposal. But this demonstration has since been furnished, with the aid of measurements obtained in a vacuum, by the Russian Lebedeff and by the Americans Nichols and Hull (1900, 1901). They have found that this pressure, the so-called

THE RADIATION PRESSURE

radiation pressure, is exactly as great as Maxwell predicted. In spite of Maxwell's great authority, astronomers quite overlooked this important law of his. Lebedeff, indeed, tried in 1892 to apply it to the tails of comets, which he regarded as gaseous; but the law is not applicable in this case. As late as the year 1900, shortly before Lebedeff was able to publish his experimental verification of this law, I attempted to prove its vast importance for the explanation of several celestial phenomena. The magnitude of the radiation pressure of the solar atmosphere must be equivalent to 2.75 milligrammes if the rays strike vertically against a black body one square centimetre in area. I also calculated the size of a spherule of the same specific gravity as water, such that the radiation pressure to which it would be exposed in the vicinity of the sun would balance the attraction by the sun. It resulted that equilibrium would be established if the diameter of the sphere were 0.0015 mm. A correction supplied by Schwarzschild showed that the calculation was only valid when the sphere completely reflects all the rays which fall upon it. If the diameter of the spherule be still smaller, the radiation pressure will prevail over the attraction, and such a sphere would be repelled by the sun. Owing to the refraction of light, this will, according to Schwarzschild, further necessitate that the circumference of the spherule should be greater than 0.3 time the wave-length of the incident rays. When the sphere becomes still smaller, gravitation will once more predominate. But spherules whose sizes are intermediate between these two limits will be repelled. It results, therefore, that molecules, which have far smaller dimensions than those mentioned, will not be repelled by the radiation pressure, and that therefore Maxwell's law does not hold for gases. When the circumference of the spherule becomes exactly equal to the wave-length of the radiation, the radiation pressure will act at its maximum, and it will then surpass gravity not less than nineteen times. These calculations apply to all spheres, totally reflecting the light, of a specific gravity like water, and to a radiation and attraction corresponding to that of the sun. Since the sunlight is not homogeneous, the maximum effect will somewhat be diminished, and it is nearly equal to ten times the gravity for spheres of a diameter of about 0.00016 mm.[1]

Before we had recourse to the radiation pressure for the explanation of the repulsion phenomena such as have been observed in the tails of comets, it was generally believed with Zöllner that the repulsion was due to electrical forces. Electricity undoubtedly plays an important part in these phenomena, as we shall see. The way in which it acts in these instances was explained by a discovery of C. T. R. Wilson in 1899. Gases can in various ways be transformed into conductors of electricity which as a rule they do not conduct. The conducting gases are said to be ionized—that is to say, they contain free ions, minute particles charged with positive or negative electricity. Gases can be ionized, among other ways, by being radiated upon with Röntgen rays, kathode rays, or ultraviolet light, as well as by strong heat. Since the light of the sun contains a great many ultra-violet radiations, it is indisputable that the masses of gases in the neighborhood of the sun (*e.g.*, probably in comets when they come near the sun) will partly be ionized.

---

[1] One centimetre of water contains 470 billions of these spheres. Such a little drop of water, again, contains 96 millions of molecules, and there are probably organisms which are smaller than these drops. Compare the experiments with ultra-microscopic organisms by E. Raehlmann, N. Gaidukow, and others.

THE RADIATION PRESSURE

and will contain both positive and negative ions. Ionized gases are endowed with the remarkable capability of condensing vapors upon themselves. Wilson showed that this property is possessed to a higher degree by the negative ions than by the positive ions (in the condensation of water vapor). If there are, therefore, water vapors in the neighborhood of the sun which can be condensed by cooling, drops of water will, in the first instance, be condensed upon the negative ions. When these drops are afterwards repelled by the radiation pressure, or when they sink, owing to gravity, as drops of rain sink in the terrestrial atmosphere, they will carry with them the charge of the negative ions, while the corresponding positive charge will remain behind in the gas or in the air. In this way the negative and positive charges will become separated from each other, and electric discharges may ensue if sufficiently large quantities of opposite electricity have been accumulated. By reason of these discharges the gases will become luminescent, although their temperature may be very low. Stark has even shown that low temperatures are favorable for the display of a strong luminosity in electric discharges.

We have stated that Kepler, as early as the beginning of the seventeenth century, came to the conclusion that the tails of comets were repelled by the sun. Newton indicated how we might, from the shape of the comets' tails, calculate their velocity. The best way, however, is to determine this velocity by direct observation. The comets' tails are not so uniform in appearance as they are generally represented in illustrations, but they often contain several luminous nuclei (Fig. 33), whose motions can be directly ascertained.

From a study of the movements of comets' tails, Olbers concluded, about the beginning of the last century, that the repulsion of the comets' tails by the sun is inversely proportional to the square of their distance—that is to say, that the force of the repulsion is subject to the same law as the force of gravitation. We can, therefore, express

Fig. 33.—Photograph of Roerdam's comet (1893 II.), suggesting several strong nuclei in the tail

the repulsion effect in units of solar gravitation, and this has generally been done. That the radiation pressure will in the same manner change with the distance is only natural. For the radiation against the same surface is

THE RADIATION PRESSURE

Fig. 34.—Photograph of Swift's comet (1892 I.)

also inversely proportional to the square of the distance from the radiating body, the sun.

In the latter part of the past century the Russian astronomer Bredichin conducted a great many measurements on the magnitude of the forces with which comets' tails are repelled by the sun. He considered himself, on the strength of these measurements, justified in dividing comets' tails into three classes. In the first class the repulsion was 19 times stronger than gravitation; in the second class, from 3.2 to 1.5 times stronger; and in the third class, from 1.3 to 1 times stronger. Still higher values have, however, been deduced for several comets.

Thus Hussey found for the comet of 1893 (Roerdam's comet, 1893 II., Fig. 33) a repulsion 37 times as strong as gravitation; and Swift's comet (1892 I.) yields the still higher value of 40.5 (Fig. 34). Some comets show several tails of different kinds, as the famous comet of Donati (Fig. 35). Its two almost straight tails would belong to the first class, and the more strongly developed and curved third tail to the second class.

Schwarzschild, as already stated, calculated that small spherules reflecting all the incident light and of the

Fig. 35.—Donati's comet at its greatest brilliancy in 1858

specific gravity of water would be repelled by the sun with a force that might balance ten times their weight. For a spherule absorbing all the light falling upon it

THE RADIATION PRESSURE

this figure would be reduced to five times the weight. The small particles of comets which, according to spectroscopic observations, probably consist of hydrocarbons are not perfectly absorbing, but they permit certain rays of the sun to pass. A closer calculation shows that in this case forces of about 3.3 times the gravity would result.

Larger spherules yield smaller values. Bredichin's second and third classes would thus be well adapted to meet the requirements which the radiation pressure demands.

It is more difficult to explain how such great forces of repulsion as those of the first group of Bredichin or of the peculiar comets of Swift and of Roerdam can occur. When a particle or drop of some hydrocarbon is exposed to powerful radiation, it may finally become so intensely heated that it will be carbonized. It will yield a spongy coal, because gases (chiefly hydrogen) will escape during the carbonization, and the particles of coal will resemble the little grains of coal-dust which fall from the smokestacks of our steamboats, and which afterwards float on the surface of the water. It is quite conceivable that such spherules of coal (consisting probably of so-called marguerites, felted or pearly structures resembling chains of bacilli) may have a specific gravity of 0.1, if we make allowance for the gases they include (compare page 106.) A light-absorbing drop of this density of 0.1 might, in the most favorable case, experience a repulsion forty times as strong as the gravitation of the sun. In this manner we can picture to ourselves the possibility of the greatest observed forces of repulsion.

The spectra of comets confirm in every respect the conclusions to which the theory of the radiation pressure leads up. They display a faint, continuous spectrum which is probably due to sunlight reflected by the small particles. Besides this, we observe, as already mentioned, a spectrum of gaseous hydrocarbons and cyanogen. These band spectra are due to electric discharges; for they are observed in comets whose distance from the sun is so great that they cannot appear luminous owing to their own high temperatures. In the tail of Swift's comet banded spectra have been observed in portions which were about five million kilometres from the nucleus. The electric discharges must chiefly be emitted from the outer parts of the tails, where, according to the laws of static electricity, the electric forces would be strongest. For this reason the larger tails of comets look as if they were enveloped in cloaks of light of a more intense luminosity.

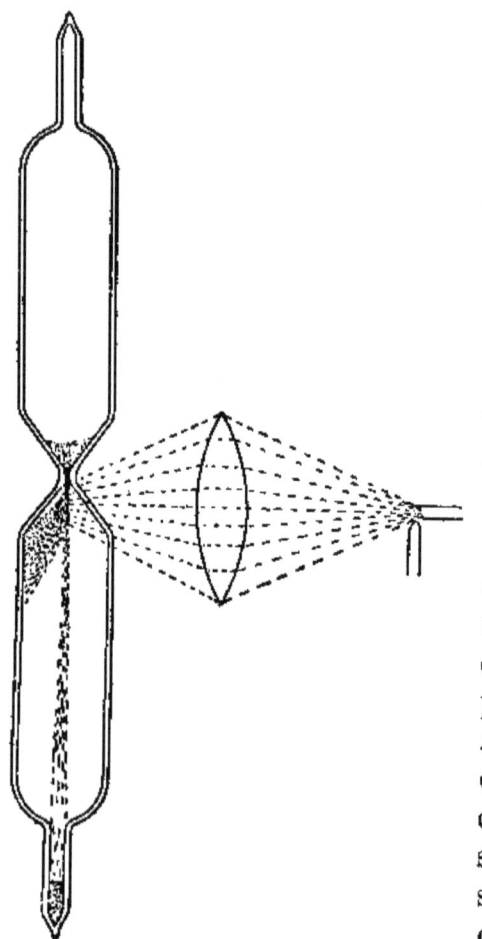

Fig. 36.—Imitation of comets' tails. Experiment by Nichols and Hull. The light of an arc-lamp is concentrated by a lens upon the stream of finely powdered particles

When a comet comes nearer to the sun, other less volatile bodies also begin to evaporate. We then find the lines of sodium and, when the comet comes very close to the sun, also the lines of iron in its spectrum. These lines are evidently produced by substances which have been evaporated from the nucleus of the comet. Like the meteorites falling upon our earth, the nucleus will consist essentially of silicates, and particularly of the silicate of sodium, and, further, of iron.

We can easily imagine how the tails of comets change in appearance. When a comet draws near to the sun, we observe that matter is ejected from that part of the nucleus which is turned towards the sun. The case is analogous to the formation of clouds in the terrestrial atmosphere on a hot summer day. The clouds are provided with a kind of hood which envelops like a thin, semi-spherical veil that side of the nucleus which turns to the sun. Sometimes we observe two or more hoods corresponding to the different layers of clouds in the terrestrial atmosphere. From the farther side of the hood matter streams away from the sun. The tails of comets are usually more highly developed when they approach the sun than when they recede from it. That may be, as has been assumed for a long time, because a large part of the hydrocarbons will become exhausted while the comet passes the sun. We have also noticed that the so-called periodical comets, which return to the sun at regular intervals, showed at every reappearance a fainter development of the tail. Comets, further, shine at their greatest brilliancy in periods of strong solar-spot activity. We may, therefore, assume that in those periods the surroundings of the sun are charged to a relatively high degree with the fine dust which can serve as a condensation nucleus for the matter of the comets' tails. It is also probable that in such periods the ionizing radiation of the sun is more pronounced than usual, owing to the simultaneous predominance of faculæ.

Nichols and Hull have attempted to imitate tails of comets. They heated the spores of the fungus *Lycoperdon bovista*, which are almost spherical and of a diameter of about 0.002 mm., up to a red glow, and they thus produced little spongy balls of carbon of an average density of 0.1. These they mixed with emery-powder and introduced them into a glass vessel resembling an hour-glass (Fig. 36) from which the air had previously been exhausted as far as possible. They then caused the powdered mass to fall in a fine stream into the lower part of the vessel while exposing it at the same time to the concentrated light of an arc-lamp. The emery particles fell perpendicularly to the bottom, while the little balls of carbon were driven aside by the radiation pressure of the light.

We also meet with the effects of the radiation pressure in the immediate neighborhood of the sun. The rectilinear extension of the corona streamers to a distance which has been known to exceed six times the solar diameter (about eight million km.) indicates that repelling forces from the sun are acting upon the fine dust. Astronomers have also compared the corona of the sun with the tails of comets, and Donitsch would class it with Bredichin's comets' tails of the second class. It is possible to calculate the mass of the corona from its radiation of heat and light. The heat radiated has been measured by Abbot. At a distance of 30,000 km. from the photosphere, the corona radiated only as little heat as a body at $-55°$ Cent. The reason is that the corona in these parts consists of an extremely attenuated mist whose actual temperature can be estimated by Stefan's law at 4300° Cent. The corona must, therefore, be so attenuated that it would only cover a 190,000th part of the sky behind it. We arrive at the same result when we calculate the amount of

THE RADIATION PRESSURE

light radiated by the corona; this radiation is of the order of that of the full moon, being sometimes smaller, sometimes greater, up to twice as great. The considerations we have been offering apply to the most intense part of the corona, the so-called inner corona. According to Turner, its light intensity outward diminishes in the inverse ratio of the sixth power of its distance from the centre of the sun. At the distance of a solar radius (690,000 km.) the light intensity would therefore be only 1.6 per cent. of the intensity near the surface of the sun.

Let us assume that the matter of the corona consists of particles of just such a size that the radiation pressure would balance their weight (other particles would be expelled from the inner corona); then we find that the weight of the whole corona of the sun would not exceed twelve million metric tons. That is not more than the weight of four hundred of our large ocean steamships (*e.g.*, the *Oceanic*), and only about as much as the quantity of coal burned on the earth within one week.

That the mass of the corona must be extremely rarefied has already been concluded, from the fact that comets have wandered through the corona without being visibly arrested in their motion. In 1843 a comet passed the sun's surface at a distance of only one-quarter the sun's radius without being disturbed in its progress. Moulton calculated that the great comet of 1881, which approached the sun within one-half its radius, did not encounter a resistance of more than one-fifty-thousandth of its mass, and that the nucleus of the comet was at least five million times denser than the matter of the corona. Newcomb has possibly expressed the degree of attenuation of the corona in a somewhat exaggerated way when he said that it contains perhaps one grain of dust per cubic kilometre (a cube whose side has a length of three-fifths of a mile).

However small the quantity of matter in the corona may be, and however unimportant a fraction of this mass may pass into the coronal rays, it is yet certain that there is a constant loss of finely divided matter from the sun. The loss, however, is not greater than the supply of matter (compare below)—namely, about 300 thousand millions of tons in a year—so that during one billion years not even one-six-thousandth of the solar mass (2×10^{27} tons) will be scattered into space. This number is very unreliable, however. We know that many meteorites fall upon the earth, partly as compact stones, partly as the finest dust of shooting-stars which flash up in the terrestrial atmosphere rapidly to be extinguished. These masses may be estimated at about 20,000 tons per year. According to this estimate, the rain of meteorites which falls upon the sun may amount to 300 thousand millions of tons in a year. All the suns have emitted matter into space for infinite ages, and it seems, therefore, a natural inference that many suns would no longer be in existence if there had not been a supply of matter to make up for this loss. The cold suns undergo relatively small losses, but receive just as large inflows of matter as the warm suns. As, now, our sun belongs to the colder type of stars, it is probable that the loss of matter from the sun has for this reason been overestimated by being presumed to be as great as the accession. The presence of dark celestial bodies may compensate for this overestimation.

Whence do the meteorites come? If they were not constantly being created, their number should have diminished in the course of ages; for they are gradually being caught up by the larger celestial bodies. It is not at all improbable that they arise from the accrescence

THE RADIATION PRESSURE

of small particles which the radiation pressure has been driving out of the sun. The chondri, which are so characteristic of meteorites, display a structure as if they had grown together out of a multitude of extremely fine grains (Fig. 37). Nordenskiöld says: "Most meteoric iron consists of an extremely delicate texture of various

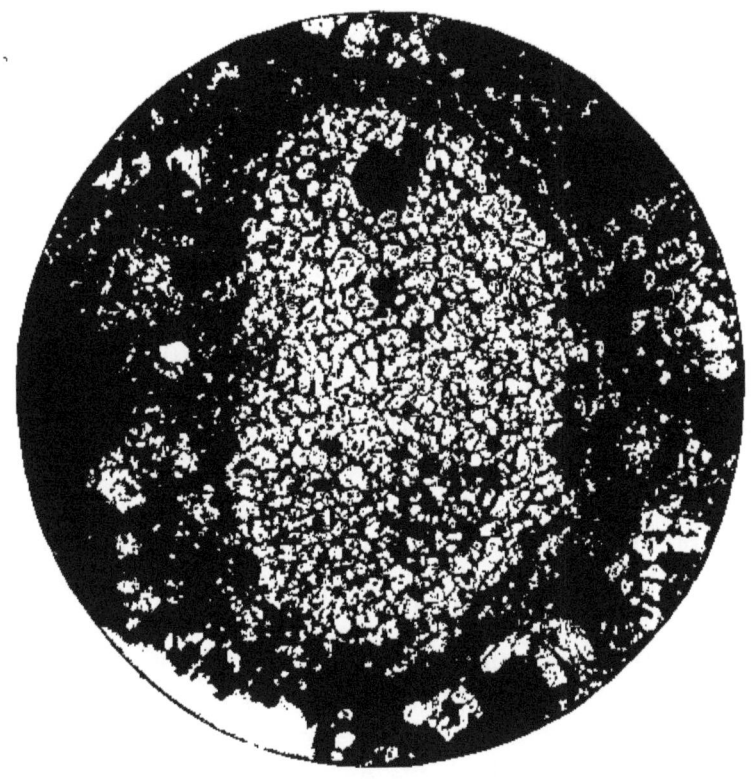

Fig. 37.—Granular chondrum from the meteorite of Sexes. Enlargement 1 : 70. After S. Tschermak

alloys of metals. This mass of meteoric iron is often so porous that it oxidizes on exposure to the air like spongy iron. The Pallas iron, when cut through with a saw, shows this property, which is so distressing for the collector. The iron of Cranbourne, of Toluca, and others—in fact, almost all the meteorites with a few exceptions—display the same texture. It all indicates that these cosmical masses of iron were built up in the universe by particle being piled upon particle, of iron, nickel, phosphorus, etc., analogous to the manner in which one atom of a metal coalesces with another atom when the metal is galvanically deposited from a solution. Most of the stony meteorites present a similar appearance. Apart from the crust of slag on the surface, the stone is often so porous and so loose that it might be used as a filtering material, and it may easily be crumbled between the fingers." When the electrically charged grains of dust coalesce, their small electrical potential (of about 0.02 volt) may increase considerably. Under the influence of ultra-violet light these masses of meteorites are discharged when they approach the sun, as Lenard has shown. Their negative charge then escapes in the shape of so-called electrons.

Since, now, the sun loses through the rays of the corona large multitudes of particles, and these particles probably carry, according to Wilson, negative electricity with them, the positive charge must remain behind in the stratum from which the coronal rays were emitted, and also on the sun itself. If this charge were sufficiently powerful, it would prevent the negatively charged particles in the corona from escaping from the sun, and all the phenomena which we have ascribed to the radiation pressure would cease. By the aid of the tenets of the modern theory of electrons, I have calculated the maximum charge that the sun could bear, if it is not to stop these phenomena. The charge would amount to two hundred thousand millions of coulombs—not by any means too large a quantity of electricity, as it would only be sufficient to decompose twenty-four tons of water.

THE RADIATION PRESSURE

By means of this positive charge the sun exerts a vast attractive power upon all negatively charged particles which come near it. We have already remarked that the grains of sun-dust which have united to form meteorites lose under the influence of ultra-violet light their charge in the shape of negative electrons, extremely minute particles, of which perhaps one thousand weigh as much as one atom of hydrogen (1 gramme of hydrogen contains about 10^{24} atoms, corresponding to 10^{27} electrons). These electrons wander about in space. When they approach a positively charged celestial body they are attracted by it with great force. If the electrons were moving with a velocity of 300 km. per second, as in Lenard's experiment, and if the sun were charged to one-tenth the maximum amount just calculated, it would be able to draw up all the electrons whose rectilinear path (so far as not curved by the sun's attraction) would lie at a distance from the sun 125 times as great as the distance between the sun and its most remote planet, Neptune, and 3800 times as great as the distance between the sun and the earth, which, after all, would only be one-sixtieth of the distance from our nearest fixed-star neighbor. The sun drains, so to speak, its surroundings of negative electricity, and this draining effect carries to the sun, as could easily be proved, a quantity of electricity which is directly dependent upon the positive charge of the sun. Thus, so far as electricity is concerned, ample provision has been made for maintaining equilibrium between the income and expenditure of the sun.

When an electrical particle enters into a magnetic field it describes a spiral about the so-called magnetic lines of force; when at a greater distance, the particles appear to move in the direction of the lines themselves. The rays of the corona emanating from the solar poles show a distinct curvature like that of the lines of force about a magnet, and for this reason the sun has been regarded as a big magnet whose magnetic poles nearly coincide with the geographical poles. The coronal rays nearer the equator likewise show this curvature (compare Fig. 30). The repelling force of the radiation pressure there is, however, at right angles to the lines of force and much stronger than the magnetic force, so that the rays of the corona are compelled to form two big streams flowing in the equatorial direction. This is especially noticeable at times of sun-spot minima. During the times of sun-spot maxima the strength of the radiation pressure of the initial velocity of the grains of dust seem to predominate so markedly that the magnetic force is relatively small.

The astronomers tell us that the sun is only a star of small light intensity compared to the prominent stars which excite our admiration. The sun further belongs to a group of relatively cold stars. We may easily imagine, therefore, that the radiation pressure in the vicinity of these larger stars will be able to move much larger masses of matter than in our solar system. If the different stars had at any time consisted of different chemical elements, this difference would have been equalized in the course of ages. The meteorites may be regarded as samples of matter collected and despatched from all possible divisions of space. Now, what bodies do we find in them?

In the comets (compare page 104) iron, sodium, carbon, hydrogen, and nitrogen (as cyanogen) play the most important part. We know, especially from the researches of Schiaparelli that meteorites often represent fragments of comets, and must therefore be related to them. Thus

THE RADIATION PRESSURE

Biela's comet, which had a period of 6.6 years, has disappeared since 1852—it had divided into two parts in 1844--1845. The comet was rediscovered in a belt of meteorites of the same period which approaches the orbit of the earth each year on November 27. Similar relations have been observed with regard to several other swarms of meteorites. We know also that the just-mentioned elements which spectrum analysis has proved to exist in comets are the main constituents of the meteorites, which, in addition, contain the metals calcium, magnesium, aluminium, nickel, cobalt, and chromium, as well as the metalloids oxygen, silicon, sulphur, phosphorus, chlorine, arsenic, argon, and helium. Their composition strongly recalls the volcanic products of so-called basic nature—that is to say, those which contain relatively large proportions of metallic oxides, and which have been thought for good reasons to hail from the deeper strata of the interior of the earth. Lockyer heated meteoric stones in the electric arc to incandescence and found their spectra to be very similar to the solar spectrum.

We therefore draw the conclusion that these messengers from other solar systems which bring us samples of their chemical elements are closely related to our sun and to the interior of our earth. That other stars and comets are essentially composed of the same elements as our sun and earth, spectrum analysis had already intimated to us. But various metalloids, like chlorine, bromine, sulphur, phosphorus, and arsenic, which are of importance for the composition of the earth, have so far not been traced in the spectra of the celestial bodies, nor in that of the sun. We find them in meteorites, however, and there is not the slightest doubt that we must likewise count them among the essential constituents of the sun and other celestial bodies. It is with difficulty, however, that the metalloids can be made to exhibit their spectra, and this is manifestly the reason why spectrum analysis has not yet succeeded in establishing their presence in the heavens. As regards the recently discovered so-called noble gases helium, argon, neon, krypton, and xenon, their presence in the chromosphere has been discovered on spectrograms taken during eclipses of the sun (Stassano). According to Mitchell, however, these statements would appear to be somewhat uncertain as to krypton and xenon.

The small particles of dust which the radiation pressure drives out into space to all possible distances from the sun and the stars may hit against one another and may accumulate to larger or smaller aggregates in the shape of cosmical dust or meteorites. These aggregates will partly fall upon other stars, planets, comets, or moons, and partly—and this in very great multitudes—they will float about in space. There they may, together with the larger dark celestial bodies, form a kind of haze, which partly hides from us the light of distant celestial bodies. Hence we do not see the whole sky covered with luminous stars, which would be the case if, as we may surmise, the stars were uniformly distributed all through the infinite space of the universe, and if there were no obstacle to their emission of light. If there were no other celestial bodies of very low temperature and very large dimensions which absorbed the heat of the bright suns, the dark celestial bodies, the meteorites, and the dark cosmical dust would soon be so strongly heated by solar radiation that they would themselves turn incandescent, and the whole dome of the sky would appear to us like one glowing vault whose hot radiation down to the earth would soon burn every living thing.

THE RADIATION PRESSURE

These other cold celestial bodies which absorb the solar rays without themselves becoming hot are known as nebulæ. More recent researches make us believe that these peculiar celestial bodies occur nearly everywhere in the sky. The wonderful mechanism which enables them to absorb heat without raising their own temperature will be explained later (in Chapter VII.). As these cold nebulæ occupy vast portions of space, most of the cosmical dust must finally, in its wanderings through infinite space, stray into them. This dust will there meet masses of gases which stop the penetration of the small corpuscles. As the dust is electrically charged (particularly with negative electricity), these charges will also be accumulated in the outer layers of the nebulæ. This will proceed until the electrical tension becomes so strong that discharges are started by the ejection of electrons. The surrounding gases will therefore be rendered luminescent, although their temperature may not much (perhaps by 50°) exceed absolute zero, $-273°$ Cent., and in this way we are enabled to observe these nebulæ. Most of the particles will be stopped before they have had time to penetrate very deeply into a nebula, and it will therefore principally be the outer portions of the nebulæ which send their light to us. That would conform to Herschel's description of planetary nebulæ, which display no greater luminosity in their centres, but which shine as if they formed hollow spherical shells of nebulous matter. It is very easy to demonstrate that only substances, such as helium and hydrogen, which are most difficult to condense, can at this low temperature exist in gaseous form to any noticeable degree. The nebulæ, therefore, shine almost exclusively in the light of these gases. There occurs in the nebulæ, in addition to these gases, a mysterious substance, nebulium, whose peculiar spectrum has not been found on the earth nor in the light of stars. Formerly the character of the nebular spectrum was explained by the assumption either that no other bodies occurred in nebulæ than the substances mentioned, or that all the other elements in them were decomposed into hydrogen—helium was not known then. The simple explanation is that only the gases of the outer layer of the nebulæ are luminous. How their interiors are constituted, we do not know.

It has been objected to the view just expressed that the whole sky should glow in a nebulous light, and that even the outer atmosphere of the earth should display such a glow. But hydrogen and helium occur only very sparely in the terrestrial atmosphere. We find, however, another light, the so-called auroral line, which may possibly be due to krypton in our atmosphere. Whichever way we turn the spectroscope on a very clear night, especially in the tropics, we observe this peculiar green line. It was formerly considered to be characteristic of the Zodiacal Light, but on a closer examination it has been traced all over the sky, even where the Zodiacal Light could not be observed. One of the objections to our view is therefore unjustified.

As regards the other objection, we have to remark that any light emission must exceed a certain minimum intensity to become visible. There may be nebulæ, and they probably constitute the majority, which we cannot observe because the number of electrically charged particles rushing into them is far too insignificant. A confirmation of this view was furnished by the flashing-up of the new star in Perseus on February 21 and 22, 1901. This star ejected two different kinds of particles, of which the one kind travelled with nearly double the velocity of the other. The accumulations of dust formed two spherical shells around the new star, corresponding in every respect to the two kinds of comets' tails of Bredichin's first and second classes, which we have sometimes observed together in the same comet (Fig. 35). When these dust particles, on their road, hit against nebular masses, the latter became luminescent, and we thereby obtained knowledge of the presence of large stellar nebulæ of whose existence we previously had not the faintest suspicion. Conditions, no doubt, are similar in other parts of the heavens where we have not so far discovered any nebulæ—we believe, because of the small number of these charged particles straying about in those parts. On the same grounds we may explain the variability of certain nebulæ which formerly appeared quite enigmatical.

V

THE SOLAR DUST IN THE ATMOSPHERE—POLAR LIGHTS AND THE VARIATIONS OF TERRESTRIAL MAGNETISM

WE have so far dwelt on the effects which the particles expelled from the sun and the stars exert on distant celestial bodies. It may be asked whether this dust does not act upon our own earth. We have already recognized the peculiar luminescence which on clear nights is diffused over the sky as a consequence of electrical discharges of this straying dust. This leads to the question whether the magnificent polar lights, which according to modern views are also caused by electric discharges in the higher strata of the atmosphere, are not produced by dust which the sun sends to us. It will, indeed, be seen that we can in this way explain quite a number of the peculiarities of these mysterious phenomena which have always excited man's imagination.

We know that meteorites and shooting-stars are rendered incandescent by the resistance which they encounter in the air at an average height of 120 km. (75 miles), sometimes of 150 and 200 km. In isolated cases meteorites are supposed to have become visible even at still greater altitudes. It would result that there must be appreciable quantities of air still at relatively high elevation, and that the atmosphere cannot be imperceptible at an altitude of less than 100 km., as was formerly assumed. Bodies smaller than the meteorites as well as the solar dust

THE SOLAR DUST IN THE ATMOSPHERE

we have spoken of—which, owing to their minuteness and to the strong cooling by heat radiation and conduction that they undergo in passing through the atmosphere, could never attain incandescence—would be stopped at greater heights. We will assume that they are arrested at a mean height of about 400 km. (250 miles).

The masses of dust which are expelled by the sun are partly uncharged, partly charged with positive or negative electricity. Only the latter can be connected with the polar lights; the former would remain dark and slowly sink through our atmosphere to the surface of the earth. They form the so-called cosmical dust, of whose great importance Nordenskiöld was so firmly convinced. He estimated that the yearly increase in the weight of the earth by the addition of the meteorites was at least ten million tons, or five hundred times more than we stated above (page 108). Like Lockyer and, in more recent days, Chamberlin, he believed that the planets were largely built up of meteorites.

The dust reaching the earth from the sun would not, were it not electrically charged, amount to more than 200 tons in a year. Although this figure may be far too low, yet the supply of matter by these means is certainly very small in comparison with the 20,000 tons which the earth receives in the shape of meteorites and shooting-stars. But owing to its extremely minute distribution, the effect of this dust is very important, and it may constitute a much greater portion of the finely distributed cosmical dust in the highest strata of the atmosphere than the dust introduced by falling meteorites and shooting-stars.

That these particles exert a noticeable influence upon terrestrial conditions, in spite of their relatively insignificant mass, is due to two causes. They are extremely minute and therefore remain suspended in our atmosphere for long periods (for more than a year in the case of the Krakatoa dust), and they are electrically charged.

In order to understand their action upon the earth, we will examine how the terrestrial conditions depend upon the position of the earth with regard to the various active portions of the sun, and upon the change of the sun itself in regard to its emission of dust particles. For this examination we have to avail ourselves of extensive statistical data; for only a long series of observations can give us a clear conception of the action of solar dust.

These particles withdraw from the sun gases which they were able to condense on their surface, and which had originally been in the chromosphere and in the corona of the sun. The most important among these gases is hydrogen; next to it come helium and the other noble gases which Ramsay has discovered in the atmosphere, in which they occur in very small quantities. As regards hydrogen, Liveing and (after him) Mitchell have maintained that it is not produced in the terrestrial atmosphere. Occasionally it is certainly found in volcanic gases. Thus hydrogen escapes, for instance, from the crater of Kilauea, on Hawaii, but it is burned at once in the atmosphere. If hydrogen were present in the atmosphere, it would gradually combine with the oxygen to water vapor; and we have to assume, therefore, that the hydrogen must be introduced into our atmosphere from another source—namely, from the sun. Mitchell finds in this view a strong support for the opinion that solar dust is always trickling down through our atmosphere.

The quantity of solar dust which reaches our atmosphere will naturally vary in proportion with the eruptive activity of the sun. The quantity of dust in the

THE SOLAR DUST IN THE ATMOSPHERE

higher strata influences the color of the light of the sun. After the eruption of the volcano Rakata on Krakatoa, in 1883, and again, though to a lesser degree, after the eruption of Mont Pelée on Martinique, red sunsets and sunrises were observed all over the globe. At the same time, another phenomenon was noticed which could be estimated quantitatively. The light of the sky is polarized with the exception of the light coming from a few particular spots. Of these spots, one called Arago's Point is situated a little above the antipode of the sun, and another, Babinet's Point, is situated above the sun. If we determine the elevation of these points above the horizon at sunset, we find in accordance with the theoretical deduction that this elevation is greater when the higher strata of the atmosphere are charged with dust (as after the eruption of Rakata) than under normal conditions. Busch, a German scientist, analyzed the mean elevation of these points (stated in degrees of arc) at sunset, and found the following peculiar numbers:

	1886	'87	'88	'89	'90	'91	'92	'93	'94	'95	Mean
Arago's Point...	20.1	19.7	18.4	17.8	17.7	20.6	19.6	20.2	**20.7**	18.8	19.4
Babinet's Point..	23.9	21.9	17.9	56.8	15.4	23.3	21.5	**24.2**	23.3	19.0	20.7
Sun-spot Number	21.1	19.1	6.7	6.1	6.5	35.6	73.8	**84.9**	78.0	63.9	40.0

There is a distinct parallelism in these series of figures. Almost simultaneously with the sun-spot maximum the height of the two so-called neutral points above the horizon attains its maximum at sunset, and the same applies to the minimum. That the phenomena in the atmosphere take place a little later than the phenomena on the sun which caused them is perhaps only natural.

When the air is rich in dust, or when it is strongly ionized by kathode rays, conditions are favorable for the formation of clouds. This can be observed, for instance, with auroral lights. They regularly give rise to a characteristic cloud formation, so much so that Adam Paulsen was able to recognize polar lights by the aid of these clouds in full daylight. Klein has compiled a table on the connection between the frequency of the higher clouds, the so-called cirrus clouds, at Cologne, and the number of sun-spots during the period 1850–1900. He demonstrates that during this half-century, which comprises more than four sun-spot periods, the sun-spot maxima fell in the years in which the greatest number of cirrus clouds had been observed. The minima of the two phenomena are likewise in agreement.

A similarly intensified formation of clouds seems also to occur on Jupiter when sun-spots are frequent. Vogel states that Jupiter at such times shines with a whiter light, while at sun-spot minima it appears of a deeper red. The deeper we are able to peep into the atmosphere of Jupiter, the more reddish it appears. During periods of strong solar activity the higher portions of Jupiter's atmosphere therefore appear to be crowded with clouds.

The discharge of the charged solar dust in our atmosphere calls forth the polar lights.

The polar lights occur, as the name indicates, most frequently in the districts about the poles of the earth. They are, however, not actually more frequent the nearer we come to the poles; but they attain a maximum of frequency in circles which enclose the magnetic and the geographical poles. The northern maximum belt passes, *via* Cape Tscheljuskin, north of Novaja Semlja, along the northwestern coast of Norway, a few degrees to the south of Iceland and Greenland, right across Hudson Bay and over the northwestern extension of Alaska. When we go to the south of this belt, the auroras, or boreal lights, diminish markedly. They are four times less frequent in Edinburgh, and fifteen times less frequent in London or New York, than in the Orkney Islands or Labrador.

Paulsen divides the auroras into two classes, which behave quite differently in several respects. The great difficulties which the solution of the problems of polar lights has so far offered seem to a large extent to be due to the fact that all polar lights were treated as being of the same kind.

The polar lights of the first class do not display any streamers. They cover a large portion of the sky in a horizontal direction. They are very quiet, and their light is strikingly constant. As a rule, they drift slowly towards the zenith, and they do not give rise to any magnetic disturbances.

These polar lights generally have the shape of an arch whose apex is situated in the direction of the magnetic meridian (Fig. 38). Sometimes several arches are grouped one above another.

Nordenskiöld observed these arches quite regularly during the polar night when he was wintering near Pitlekaj, in the neighborhood of Bering Sound. Adam Paulsen has often seen them on Iceland and Greenland, which are situated within the maximum belt spoken of, where northern lights are very common. Occasionally, auroras are also seen farther from the poles, as circular arches of a milky white, which may be quite high in the heavens.

Sometimes we perceive in the arctic regions that large areas of the heavens are covered by a diffused light which might best be compared to a luminous, transparent cloud; the darker portions in it probably appear dark by contrast. This phenomenon was frequently observed dur-

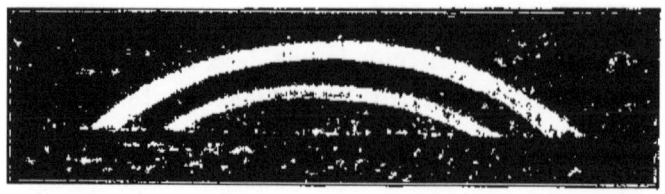
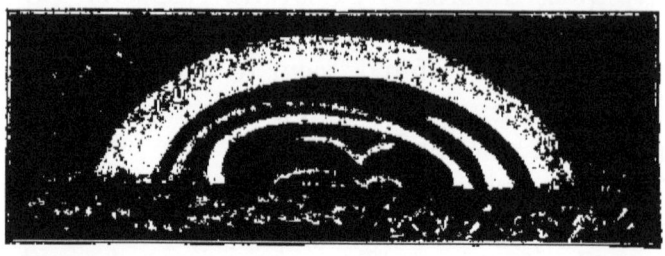
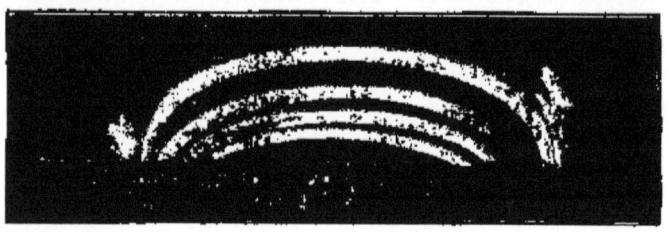

Fig. 38.—Arch-shaped auroræ borealis, observed by Nordenskiöld during the wintering of the *Vega* in Bering Strait, 1879

ing the Swedish expedition of 1882-1883, near Cape Thordsen.

Masses of light at so low a level that the rocks behind them are obscured have frequently been observed to float in the air, especially in the arctic districts. Thus Lemström saw an aurora on the island of Spitzbergen in front of a wall of rock only 300 m. (1000 ft.) in height. In northern Finland he observed the auroral line in the light of the air in front of a black cloth only a few metres distant. Adam Paulsen counts these phenomena also as polar lights of the first class, and he regards them as phosphorescent clouds which have been carried down by convection currents to an unusually low level of our atmosphere.

Polar lights of the second class are distinguished by the characteristic auroral rays or streamers. Sometimes these streamers are quite separated from one another (see Fig. 39); as a rule they melt into one another, especially below, so as to form draperies which are so easily moved and unsteady that they appear to flutter in the

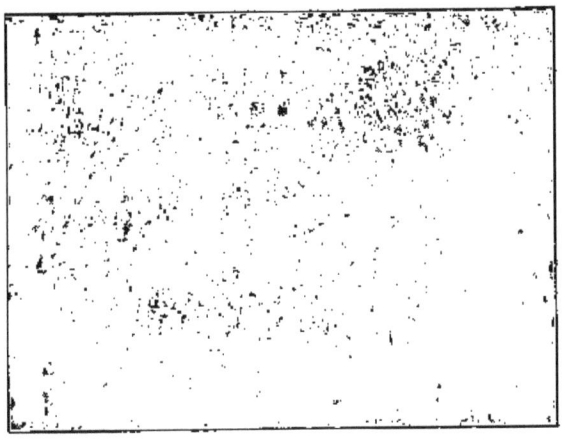

Fig. 39.—Aurora borealis, with radial streamers

wind (Fig. 41.) The streamers run very approximately in the direction of the inclination (magnetic dip) needle, and when they are fully developed around the celestial dome their point of convergence is distinctly discernible in the so-called corona (Fig. 40). When the light is at its greatest intensity the aurora is traversed by numerous waves of light.

The draperies are very thin. Paulsen watched them sometimes drifting over his head in Greenland. The draperies then appeared foreshortened, in the shape of striæ or ribbons of light in convolutions. These polar lights influence the magnetic needle. When they pass the zenith their influence changes sign, so that the deviation of the magnetic needle changes from east to west when the rib-

Fig. 40.—Aurora with corona, observed by Gyllenskiöld on Spitzbergen, 1883

bon is moving from north to south. Paulsen therefore concluded that negative electricity (kathode rays) was moving downward in these rays. These polar lights correspond to violent displacements of negative electricity, while polar lights of the first class appear to consist of a phosphorescent matter which is not in strong agitation. The streamers may penetrate down into rather low atmospheric strata, at least in districts which are near the maximum belt of the northern lights. Thus Parry observed at Port Bowen an auroral streamer in front of a cliff only 214 m. (700 ft.) in height.

Polar lights of the first order may pass into those of the second order, and *vice versa*. We frequently see rays suddenly flash out from the arch of the aurora, mostly downward, but, when the display is very intense, also upward. On the other hand, the violent agitation of a "drapery light" may cease, and may give way to a diffused, steady glow in the sky. The polar light of the first class is chiefly observed in the arctic regions. To it corresponds, in districts farther removed from the pole, the diffused light which appears to be spread uniformly over the heavens and which gives the auroral line.

The usually observed polar lights (speaking not only of those seen on arctic expeditions) belong to the second class, which comprises also all those included in the subjoined statistics, with the exception of the auroral displays reported from Iceland and Greenland. While the

Fig. 41.—Polar-light draperies, observed in Finnmarken, northern Norway

streamer lights distinctly conform to the 11.1 years' period, and become more frequent at times of sun-spot maxima, this is not the case, according to Tromholt, with the auroras of Iceland and Greenland. Their frequency, on the contrary, seems to be rather independent of the sun-spot frequency. Not rarely auroral maxima corresponding to sun-spot maxima are subdivided into two by a secondary minimum. This phenomenon is most evident in the polar regions, but it can also be traced in the statistics from Scandinavia and from other countries.

Better to understand the nature of auroras, we will consider the sun's corona during the time of a minimum year, taking as an example the year 1900 (compare Fig. 30). The rays of the corona in the neighborhood of the poles of the sun are laterally deflected by the action of the magnetic lines of force of the sun. The small, negatively charged particles have evidently only a low velocity, so that they move quite close to the lines of force in the neighborhood of the solar poles and are concentrated near the equator. There the lines of force are less crowded —that is to say, the magnetic forces are weaker—and the solar dust can therefore be ejected by the radiation pressure and will accumulate to a large disk expanding in the equatorial plane. To us this disk appears like two large streams of rays which project in the direction of the solar equator. Part of this solar dust will come near the earth and be deflected by the magnetic lines of force of the earth; it will hence be divided into two streams which are directed towards the two terrestrial magnetic poles. These poles are situated below the earth's crust, and therefore not all the rays will be concentrated towards the apparent position of the magnetic poles upon the surface of the earth. It is to be expected that the negatively charged particles coming from the sun will chiefly drift towards that district which is situated somewhat to the south of the magnetic north pole, when it is noon at this pole. When it is midnight at the magnetic pole, most of the negatively charged particles will be caught by the lines of force before they pass the geographical north pole, and the maximum belt of the auroras will for this reason surround the magnetic and the geographical poles, as has already been pointed out (compare page 122). The negatively charged solar dust will thus be concentrated in two rings above the maximum belts of the polar lights. Where the dust collides with molecules of the air, it will produce a phosphorescent glow, as if these molecules were hit by the electrically charged particles of radium. This phosphorescent glow rises in the shape of a luminous arch to a height of about 400 km. (250 miles)—according to Paulsen—and the apex of this arch will in every part seem to lie in the direction where the maximum belt is nearest to the station of the observer. That will fairly coincide with the direction of the magnetic needle.

The solar corona of a sun-spot maximum year is of a very different appearance (Fig. 31). The streamers radiate straight from the sun in almost all directions; and if there be some privileged directions, it will be those above the sun-spot belts. The velocity of the solar dust is evidently so great that the streamers are no longer visibly deflected by the magnetic lines of force of the sun. Nor is this charged dust influenced to any noticeable degree by the magnetic lines of force of the earth. It will in the main fall straight down in that part of the atmosphere in which the radiation is most intense. As these "hard" rays of the sun[1] seem to issue from the faculæ of the sun which are most frequent in maximum sun-spot

[1] The designations "hard" and "soft" streams of solar dust correspond to the terms used with regard to kathode rays. The soft rays have a smaller velocity, and are therefore more strongly deflected by external forces, as, for instance, magnetic forces.

years, some polar lights will also be seen in districts which are far removed from the maximum belt of the auroras, especially when the number of sun-spots is large. The opposite relation holds for the "soft" streams of solar dust which fall near the maximum belt of the polar lights. These streams occur most frequently with low sun-spot frequency, as we know from observations of the solar corona. Possibly they are carried along by the stream of harder rays in maximum years. The polar lights corresponding to these rays therefore attain their maximum with few sun-spots. Hard and soft dust streams occur, of course, simultaneously; but the former predominate in maximum sun-spot years, the latter in minimum years.

That the periodicity of the polar lights in regions without the maximum belt follows very closely the periodicity of the sun-spots was shown by Fritz as early as 1863. The length of the period varies between 7 and 16 years, the average being 11.1 years. The years of maxima and minima for sun-spots and for northern auroras are the following:

MAXIMUM YEARS

Sun-spots........1728 '39 '50 '62 '70 '78 '88 1804 '16 '30
                 1837 '48 '60 '71 '83 '93 1905
Northern lights...1730 '41 '49 '61 '73 '78 '88 1805 '19 '30
                  1840 '50 '62 '71 '82 '93 1905

MINIMUM YEARS

Sun-spots........1734 '45 '55 '67 '76 '85 '98 1811 '23 '34
                 1841 '56 '67 '78 '89 1900
Northern lights...1735 '44 '55 '66 '75 '83 '99 1811 '22 '34
                  1844 '56 '66 '78 '89 1900

There are, in addition, as De Mairan proved in his classical memoir of the year 1746, longer periods common to both the number of sun-spots and the number of auroras. According to Hansky, the length of this period is 72 years; according to Schuster, 33 years. Very pronounced maxima occurred at the beginning and the end of the eighteenth century, the last in the year 1788; afterwards auroras became very rare in the years 1800–1830, just as in the middle of the eighteenth century. In 1850, and particularly in 1871, there were strong maxima; they have been absent since then.

The estimates of the heights of the polar lights vary very considerably. The height seems to be the greater, on the whole, the nearer the point of observation is to the equator, which would well agree with the slight deflection of the kathode rays towards the surface of the earth in regions which are farther removed from the pole. Gyllenskiöld found on Spitzbergen a mean height of 55 km.; Bravais, in northern Norway, 100 to 200 km; De Mairan, in central Europe, 900 km.; Galle, again, 300 km. In Greenland, Paulsen observed northern lights at very low levels. In Iceland he fixed the apex of the northern arch which may be considered as a point where the charged particles from the sun are discharged into the air at about 400 km. Not much reliance can be placed upon the earlier determinations; but the heights given conform approximately to the order of magnitude which we may deduce from the height at which the solar dust will be stopped by the terrestrial atmosphere.

The polar lights possess, further, a pronounced yearly periodicity which is easily explicable by the aid of the solar dust theory. We have seen that sun-spots are rarely observed near the solar equator, and the same applies to solar faculae. They rapidly increase in frequency with higher latitudes of the sun, and their maximum occurs at latitudes of about fifteen degrees. The equatorial plane of the sun is inclined by about seven degrees towards the plane of the earth's orbit. The earth is in the equatorial plane of the sun on December 6th and June 4th, and most distant from it three months later. We may, therefore, expect that the smallest number of solar-dust particles will fall on the earth when the earth is in the equator of the sun—that is, in December and June—and the greatest number in March and September. These relations are somewhat disturbed by the twilight, which interferes with the observation of auroras in the bright summer nights of the arctic region, while the dark nights of the winter favor the observation of these phenomena. The distribution of the polar lights over the different seasons of the year will become clear from the subjoined table compiled by Ekholm and myself:

	Sweden (1883-96)	Norway (1861-95)	Iceland and Greenland (1872-92)	United States (1871-93)	Southern aurorae (1856-94)
January	1056	**251**	804	1005	56
February	1173	331	734	1455	126
March	**1312**	**335**	613	1396	**183**
April	568	90	128	**1724**	148
May	170	6	1	1270	54
June	10	0	0	1061	40
July	54	0	0	1233	35
August	101	18	40	1210	75
September	1055	209	455	**1735**	120
October	**1114**	**353**	716	1630	**192**
November	1077	326	811	1240	112
December	**940**	260	**863**	912	81
Average number	727	181	430	1322	102

In zones where the difference between the lengths of day and night of the different seasons is not very great, as in the United States, and in districts in which the southern light is observed (about latitude 40° S.), the chief minimum falls in winter: on the northern hemisphere, in December; on the southern hemisphere, in June or July. A less pronounced minimum occurring in the summer. Twice in the course of the year the earth passes through the plane of the solar equator. During these periods a minimum of solar dust trickles down upon the earth, and that period is characterized by a larger number of polar lights which is distinguished by a higher elevation of the sun above the horizon. We may expect this; for most solar dust will fall upon that portion of the earth over which the sun is highest at noon. The two maxima of March or April and of September or October, when the earth is at its greatest distance from the plane of the solar equator, are strongly marked in all the series, except in those for the polar districts Iceland and Greenland. There the auroral frequency is solely dependent upon the intensity of the twilight, so that we find a single maximum in December and the corresponding minimum in June. More recent statistics (1891-1903) indicate, however, a minimum in December. For the same reason the summer minimum in countries of high latitudes, like Sweden and Norway, is very much accentuated.

Similar reasons render it difficult for most localities to indicate the daily periodicity of the polar lights. Most of the solar dust falls about noon, and most polar lights should therefore be counted a few hours after noon, just as the highest temperature of the day is reached a little after noon. On account of the intense sunlight, however, this maximum can only be established in the wintry night of the polar regions, and even there only when a correction has been made for the disturbing effect of the twilight. In this way Gyllenskiöld found a northern-light maximum at 2.40 P.M. for Cape Thordsen, on Spitzbergen, the corresponding minimum being at 7.40 A.M. In other localities we can only ascertain that the polar lights are more intense and more frequent before than after midnight. In central Europe the maximum occurs at about 9 P.M.; in Sweden and Norway (in latitude 60° N.), half an hour or an hour later.

A few other periods, approximately of the length of a month, have been suggested with regard to polar lights. A period lasting 25.93 days predominates in the southern lights, where the maximum exceeds the average by 44 per cent. For the northern lights in Norway the corresponding excess percentage is 23; for Sweden, only 11.[1]

The same period of nearly twenty-six days had already been pointed out for a long series of other especially magnetic phenomena which, as we shall see, are very closely connected with auroras, and it had also been found in the frequency of thunder-storms and in the variations of the barometer. This periodicity has often been thought to be connected with the axial rotation of the sun. The Austrian scientist Hornstein has even gone so far as to propose that the length of this period should be carefully determined, "because it would give a more accurate value for the rotation of the sun than the direct determinations." We know now that the length of the solar revolution is different for different solar altitudes, a circumstance with which observations of sun-spot movements at different latitudes had already made Carrington and Spörer familliar, but which was not safely established before Dunér's spectroscopical determination of the movement of the solar

---

[1] The reason is that in the southern district only very few, and chiefly the most intense, auroras are recorded. If we observe very assiduously in a large country, and conduct the observations at different spots, we shall find polar light almost every night. This consideration partly wipes out the just-mentioned differences.

VARIATIONS OF TERRESTRIAL MAGNETISM

photosphere. Dunér found the following sidereal revolutions for different latitudes of the sun to which the subjoined synodical revolution would correspond. (By sidereal revolution of a point on the sun we understand the time which elapses between the two moments when a certain star passes, on two consecutive occasions, through the meridian plane of the point—that is to say, through a plane laid through the poles of the sun and the point in question. The synodical revolution is determined by the passage of the earth through this meridian. On account of the proper motion of the earth the synodical period is longer than the sidereal period.)

Latitude on the sun (degrees)...	0	15	30	45	60	75
Sidereal revolution (days)	25.4	26.4	27.6	30.0	33.9	38.5
Synodical revolution (days)	27.3	28.5	29.9	32.7	37.4	43.0

That the periods of rotation of the solar photosphere, and, in a similar way, the periods of the spots, the faculæ, and the prominences, should become so considerably longer with increasing latitudes is one of the most mysterious problems of the physics of the sun. Something similar applies to the clouds of Jupiter, but the difference in that case is much smaller—only about one per cent. The clouds of our atmosphere behave quite differently, a fact which is explained by our atmospheric circulation.[1]

In our case, of course, the position of the sun with regard to the earth—that is to say, the synodical period—can

[1] The very highest strata of our atmosphere (at levels of from 20 to 80 km., 15 to 50 miles) may perhaps form an exception. The luminous clouds which were observed in the years 1883–1892 at Berlin (after the eruption of Krakatoa), and which were floating at a very high level, showed a drift with regard to the surface of the earth opposite to the drift of the cirrus clouds, which are directed eastward.

alone be of importance. We recognize that the period of 25.93 days does not at all agree with any period of the solar photosphere. The solar equatorial zone differs least, and it would be appropriate to reckon with this period, since the earth never moves very far from the plane of the solar equator, and returns to that plane, at any rate, twice in the course of a year.

But there is another peculiarity. The higher a point is situated in the atmosphere of the sun, the shorter is its period. Thus the synodical period of the faculæ near the equator is on an average 26.06, the period of the spots 26.82, of the photosphere 27.3 days. Faculæ situated at higher levels revolve still more rapidly, and we are thus driven to the conclusion that the period to which we have alluded agrees with the period of the faculæ situated at higher levels in the equatorial zone of the sun, and is probably dependent upon them. That would conform to our ideas concerning the physics of the sun. For the faculæ are produced in the ascending currents of gas and at rather lower levels than the spherules which are expelled by the radiation pressure. This radiation pressure is strongest just in the neighborhood of the faculæ.

For the same reason the repulsion of the solar dust becomes particularly powerful when the faculæ are strongly developed—that is to say, just in the time of pronounced eruptive activity of the sun which is characterized by many sun-spots.

We must thus imagine that the radiation of the sun will be stronger in times of strong eruptive activity than during the days of low sun-spot frequencies. Direct observations of the intensity of the solar radiation which have been made by Saveljeff in Kieff confirm this assumption. It must be pointed out, however, that another phenomenon investigated by Köppen seems to contradict this conclusion. Köppen ascertained that in our tropics the temperature was by 0.32° Cent. (nearly 0.6° F.) lower during sun-spot maxima than the average, and that five years later, a year before the sun-spot minimum, it reached its maximum value of 0.41° Cent. (0.7° F.) above the average. A similar peculiarity can be traced to other zones, but on account of the less steady climates it is much less marked there than in the tropics. A French physicist, Nordmann, has fully confirmed the observations of Köppen. On the other hand, Very, an American astronomer, has found that the temperature in very dry (desert) districts of the tropics (near Port Darwin, 12° 28' S., and near Alice Springs, 23° 38' S., both in Australia) is higher at sun-spot maxima than at minima; but Very was in this research guided merely by the records of maximum and minimum thermometers. From Very's investigation it would appear that the solar radiation is really more intense with larger sun-spot numbers.[1] This, it must be remarked, is only noticeable in exceedingly dry districts in which there is no cloud formation worth mentioning. In other districts the cloud formation which accompanies sun-spot maxima interferes with the simplicity of the phenomena. The cooling effect of the clouds seems in these cases by far to surpass the direct heating effect of the solar rays, and in this manner Köppen's conclusion would become explicable. If we could observe the temperatures of the atmospheric strata above the clouds, their variation would no doubt be in the same degree as that in the desert.

Finally, we have to note another period in the phe-

[1] According to Memery (*Bull. Soc. Astr.*, March 7, 1906, p. 168) an instantaneous rise of temperature is observed immediately when a sun-spot is first seen, and the temperature sinks again when the sun-spot disappears.

nomena of the polar lights—the so-called tropical month, whose length is 27.3 days. The nature of this period is little understood. It is possibly connected with the

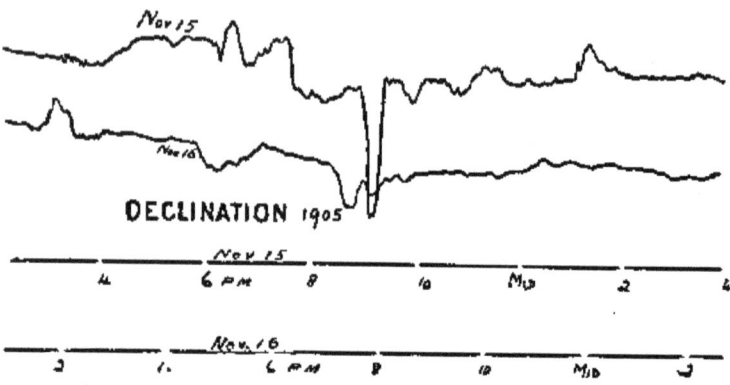

Fig. 12.—Curve of magnetic declination at Kew, near London, on November 15 and 16, 1905. The violent disturbance of November 15, 9 P.M., corresponds to the maximum intensity of the aurora. Compare the following figure

electric charge of the moon. The peculiarity of this period is that it acts in an opposite way in the northern and southern hemispheres. When the moon is above the horizon, it seems to prevent the formation of polar lights; but for this case the difficulties of observation caused by the moonlight must, of course, be taken into consideration.

Celsius and Hiorter observed in 1741 that the polar lights exercise an influence on the magnetic needle. From this circumstance we have drawn the conclusion that the polar lights are in some way due to electric discharges which act upon the magnetic needle. These magnetic effects, the disturbances of the otherwise steady position of the magnetic needle, are not influenced by the light of the sun and moon, and can therefore be

studied to greater advantage than auroras. We have already pointed out that it is only the aurora of the radial, streamer type which exerts this magnetic influence (compare Figs. 42 and 43).
These magnetic variations show exactly the same periods as the northern lights and the sun-spots. As regards, first, the long period of 11.1 years, our observations prove that the so-called magnetic disturbances of the position of the magnetic needle faithfully reflect the variations in the sun-spots. This connection was discovered in 1852 by Sabine in England, by Wolf in Switzerland, and by Gautier in France. Even the more irregular diurnal variations in the magnetic elements are subjected to a solar period. The magnetic needle points in our districts with its north end towards the north—not exactly, though, being deflected towards the west. This

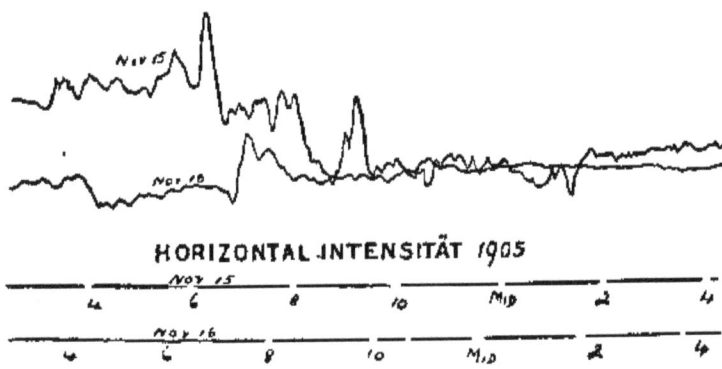

Fig. 43.—Curve of horizontal intensity at Kew on November 15 and 16, 1905. On November 15 a magnificent aurora was observed in Galicia, Germany, France, Norway, England, Ireland, and Nova Scotia, with a maximum at 9 P.M. The polar light was unusually brilliant as early as 6 P.M.

western deviation or declination is greatest soon after noon, about one o'clock, and this diurnal change is greater in summer than in winter, and the fluctuation of the position of the magnetic needle greater in daytime than at night-time. It is, therefore, manifest that we have to deal with some solar effect. This becomes still more distinct when we study the change with reference to the daily variation in the number of sun-spots. In the subjoined table the variation in the declination has been compiled for Prague for the years 1856 to 1889. Only years with maxima and minima of sun-spots and of magnetic variations have been noticed in this table:

	1856	1860	1867	1871	1879	1884	1889
Sun-spot number..	4.3	95.7	7.3	139.1	3.4	63.7	6.3

DAILY VARIATIONS IN DECLINATION

	1856	1859	1867	1871	1878	1883	1889
Observed...	5.98	10.36	6.95	11.43	5.65	8.34	5.99
Calculated.	6.08	10.20	6.22	12.15	6.04	8.76	6.17

We see that the maxima and minima years of the two phenomena very nearly coincide. The accord is so evident that we may calculate the diurnal variation as proportional to the increase in the number of sun-spots. This is shown by the two last lines of the table.

The yearly variation is again exactly the same as that of polar lights, as the following table indicates, in which the disturbances of magnetic declination, horizontal intensity, and vertical intensity are compiled for Toronto, Canada; for comparison the means of these three magnitudes are added for Greenwich. The unit of this table is the average annual variation:

TORONTO

	Jan.	Feb.	Mar.	April	May	June	July	Aug.	Sept.	Oct.	Nov.	Dec.
Declination.	0.57	0.84	1.11	1.42	0.98	0.53	0.94	1.16	1.62	1.31	0.78	0.76
Horizontal..	0.56	0.94	0.94	1.50	0.90	0.36	0.61	0.75	1.71	1.48	0.98	0.58
Vertical....	0.57	0.74	1.08	1.49	1.12	0.50	0.71	1.08	1.61	1.29	0.76	0.61

GREENWICH

	Jan.	Feb.	Mar.	April	May	June	July	Aug.	Sept.	Oct.	Nov.	Dec.
Means..	0.93	1.23	1.22	1.09	0.81	0.71	0.81	0.90	1.15	1.18	1.02	0.83

VARIATIONS OF TERRESTRIAL MAGNETISM

The daily variation of the disturbances has been analyzed by Van Bemmelen for the period 1882-1893 and for the observatory of Batavia, on Java. The maximum occurs there about 1 P.M., and is about 1.86 times as great as the average value for the day. The minimum of 0.48 occurs at 11 P.M. Between 8 P.M. and 3 A.M. the disturbances are almost as rare as about 11 o'clock at night.

The variation is greatest with that declination which has its maximum of 3.26 at 12 M., and its minimum of 0.14 at 11 P.M.

The period of almost 26 days first investigated by Hornstein has also been refound in the magnetic variations and disturbances by Broun, Liznár, and C. A. Müller. It must be added, however, that Schuster does not consider these data as in any way conclusive.

The moon has also a slight influence upon the magnetic needle, as Kreil proved as early as 1841. The effect is in a different sign in the northern and southern hemispheres, and may be likened to a tidal phenomenon.

The ultra-violet rays of the sun are strongly absorbed by the atmosphere, and they cause an ionization of the molecules of the air. This ionization is, on the whole, more marked at higher altitudes. The ascending air currents carry with them water vapor which is condensed on the ions, particularly on the negative ions. In this way most clouds become negatively charged; this interesting fact—*i. e.*, that they are more frequently charged with negative than with positive electricity—was first proved by Franklin in his kite experiments. When the rain-drops have fallen, the air above remains positively charged; this has been observed during balloon ascensions. The clouds which are formed at high levels are most strongly charged; for this reason thunder-storms over land occur mostly in the summer-time. The thunder-storms also show the 26-day period, as Bezold has proved for southern Germany, and Ekholm and myself have shown for Sweden.

A vast amount of material concerning these questions and magnetic phenomena in particular has been collected by the various meteorological observatories and is awaiting analysis.

Although some observers like Sidgreaves question the correlation of sun-spots and polar lights or magnetic disturbances, because strong spots have been seen on the disk of the sun without any magnetic disturbances having been noticed, yet the view predominates that the magnetic disturbances are caused by sun-spots when the sun-spots cross the central meridian of the sun which is opposite the earth. Thus Maunder observed a magnetic storm and a northern light succeeding the passage of a large sun-spot through the central solar meridian on the 8th to the 10th of September, 1898. The magnetic effect attained its maximum about twenty-one hours after the passage through the meridian.

Similarly Riccò established in ten instances, in which exact determination was possible, a time interval of 45.5 hours on an average between the meridian passage of a spot and the maximum magnetic effect. Riccò also submitted to an analysis the data which Ellis had collected and which Maunder had investigated. He found for these instances, on an average, almost exactly the same numbers, the time interval being 42.5 hours. That would correspond to a mean velocity of the solar dust of from 910 to 980 km. per second. On the other hand, we have no difficulty at all in calculating the time which a spherule of a diameter of 0.00016 mm. (those particles travel fastest) and of the specific gravity of water would need in order to reach the earth, under the influence of solar gravitation and of a mechanical radiation pressure 2.5 times as large from the outside of the sun. The time found, 56.1 hours, corresponds to a mean velocity of 740 km. per second. In order that the solar dust may move with the velocity calculated by Riccò, its specific gravity should be less than 1—*viz.*, 0.66 and 0.57. This value looks by no means improbable, when we assume that the spherules consist of hydro-carbons saturated with hydrogen, helium, and other noble gases. We should also arrive at larger velocities for the solar dust, as has already been pointed out with regard to the tails of comets, when we presume that the particles consist of felted marguerites of carbon or silicates, or of iron—materials which we regard as the main constituents of meteorites.

It should, perhaps, be mentioned that the most intense spectrum line of the polar lights has been found to be characteristic of the noble gas krypton. As this gas is found only in very small quantities in our atmosphere, it does not appear improbable that it has been carried to us together with the solar dust, and that its spectrum becomes perceptible during the discharge phenomena. The other spectrum lines of the polar lights belong to the spectra of nitrogen, argon, and of the other noble gases. The volumes of noble gases which are brought into our terrestrial atmosphere in this manner would in any case be exceedingly small.

The electrical phenomena of our terrestrial atmosphere indirectly possess considerable importance for organic life and, consequently, for human beings. By the electrical discharges part of the nitrogen is made to combine with the oxygen and hydrogen (liberated by the electric decomposition of water vapor) of our air, and it thus forms the ammonia compounds, as well as the nitrates and nitrites, which are so essential to vegetable growth. The ammonia compounds which play a most important part in the temperate zones appear especially to be formed by the so-called silent discharges which we connect with auroras. The oxygen compounds of nitrogen, on the other hand, seem to be chiefly the products of the violent thunder-storms of the tropics. The rains carry these compounds down into the soil, where they fertilize the plants.

The supply of nitrogen thus fixed amounts in the course of a year to about 1.25 gramme per square metre in Europe, and to almost fourfold that figure in the tropics. If we accept three grammes as the average number for the whole firm land of the earth, that would mean 3 tons per square kilometre, and about 400 million tons per year for the whole firm land of 136 million square kilometres. A very small portion of this fixed nitrogen, possibly one-twentieth, falls on cultivated soil; a larger portion will help to stimulate plant growth in the forests and on the steppes. We may mention, for comparison, that the nitrogen contained in the saltpetre which the mines of Chili yield to us has risen from 50,000 tons in 1880 to 120,000 tons in 1890, to 210,000 tons in 1900, and to 260,000 tons in 1905. The nitrogen produced in the shape of ammonium salts (sulphate) by the gas-works of Europe amounts to about one-quarter of the last-mentioned figure. To this figure we have, of course, to add the production of coal-gas ammonia in the United States and elsewhere. Yet even allowing for this item, we shall find that the artificial supply of combined nitrogen on the earth does not represent more than about one-thousandth of the natural supply.

As the nitrogen contents of the air may be estimated

at 3980 billion tons, we recognize that only one part in three millions of the nitrogen of the atmosphere is every year fixed by electric discharges, presuming that the nitrogen supply to the sea is as great per square kilometre as to the land. The nitrogen thus fixed benefits the plants of the sea and of the land, and passes back into the atmosphere or into the water during the life of the plants or during their decay. Water absorbs some nitrogen, and equilibrium between the nitrogen contents of the atmosphere and of the sea is thus maintained. Hence we need not fear any noteworthy depletion of the nitrogen contents of the air. This conclusion is in accord with the fact that no notable accumulation of fixed nitrogen appears to have taken place in the solid and liquid constituents of the earth.

The reader may remember (compare page 57) that during the annual cycle of vegetation not less than one-fiftieth of the atmospheric contents in the carbon dioxide is absorbed. Since oxygen is formed from this carbon dioxide, and since the air contains about seven hundred times as many parts per volume of oxygen as of carbon dioxide, the exchange of atmospheric oxygen is about one part in thirty-five thousand. In other words, the oxygen of the air participates about one hundred times more energetically in the processes of vegetation than the nitrogen, and this is in accordance with the general high chemical activity of oxygen.

Before we close this chapter we will briefly refer to the peculiar phenomenon known as the Zodiacal Light, which can be seen in the tropics almost any clear night for a few hours after or before sunset or sunrise. In our latitudes the light is rarely visible, and is best seen about the periods of the vernal and autumnal equinoxes. The phenomenon is generally described as a luminous cone

WORLDS IN THE MAKING

whose basis lies on the horizon, and whose middle line coincides with the zodiac. Hence the name. According to Wright and Liais, its spectrum is continuous. It is stated that the Zodiacal Light is as strong in the tropics as that of the Milky Way.

There can be no doubt that this glow is due to dust particles illuminated by the sun. It has therefore been assumed that this dust is floating about the sun in a ring,

Fig. 44.—Zodiacal Light in the tropics

and that it represents the rest of that primeval nebula out of which the solar system has been condensed, according to the theory of Kant and Laplace (compare Chapter VII.). Sometimes a fairly luminous band seems to shoot out from the apex of the cone of the Zodiacal Light and to cross the sky in the plane of the ecliptic. In that part of the sky which is just opposite the sun this band expands to a larger, diffused, not well-defined spot of light covering about $12°$ of arc in latitude and $90°$ in longitude. This luminescence is called the counter-glow (Gegenschein), and was first described by Pezenas in 1780.

The most probable view concerning the nature of this counter-glow is that it is caused by small particles of meteorites or dust which fall towards the sun. Like the position of the corona of the aurora, the position of this counter-glow seems to be an effect of perspective; the orbits of the little particles are directed towards the sun, and they therefore appear to radiate from a point opposite to it.

We know very little about this phenomenon. Even the position of the Zodiacal Light along the zodiac which has given rise to its name has been questioned, and it would appear from recent investigations that the glow is situated in the plane of the solar equator. However that may be, the view is very generally held that the glow is due to particles which come from the sun or enter into it. We have already adduced arguments to prove that the mass of solar dust cannot be unimportant; this dust may therefore be the cause of the phenomenon which we have just been discussing.

VI

END OF THE SUN—ORIGIN OF NEBULÆ

We have seen that the sun is dissipating and wasting almost inconceivable amounts of heat every year: $3.8 . 10^{33}$ gramme-calories, corresponding to 2 gramme-calories for each gramme of its mass. We have also obtained an idea as to how the enormous storage of heat energy in the sun may endure this loss for ages. Finally, however, the time must come when the sun will cool down and when it will cover itself with a solid crust, as the earth and the other planets — so far, probably, in a gaseous state—have done long since or will do before long. No living being will be able to watch this extinction of the sun despairingly from one of the wandering planets; for, in spite of all our inventions, all life will long before have ceased on the satellites of the sun for want of heat and light.

The further development of the cold sun will recall the actual progress of our earth, except in so far as the sun will have no life-spending, central source of light and heat near it. In the beginning the thin, solid crust will again and again be burst by gases, and streams of lava will rush out from the interior of the sun. After a while these powerful discharges will stop, the lava will freeze, and the fragments will close up more firmly than before. Only on some of the old fissures volcanoes will rise and allow the gases to escape from the interior—water vapor

END OF THE SUN

and, to a less extent, carbonic acid, liberated by the cooling.

Then water will be condensed. Oceans will flood the sun, and for a short period it will resemble the earth in its present condition, though with the one important difference. The extinct sun, unlike our earth, will not receive life-giving heat from the outside, excepting the small amount of radiation from universal space and the heat generated by the fall of meteorites. The temperature fall will therefore be rapid, and the vanishing clouds of the attenuated atmosphere will not long check radiation. The ocean will become covered with a crust of ice. Then the carbonic acid will commence to condense, and will be precipitated as a light snow in the solar atmosphere. Finally, at a temperature of about $-200°$ Cent., the gases of the atmosphere will be condensed, and new oceans, now principally of nitrogen, will be produced. Let the temperature sink another $20°$, and the energy of the inrushing meteorites will just suffice to balance a further loss of heat by radiation. The solar atmosphere will then consist essentially of helium and hydrogen — the two gases which are most difficult to condense—and of some nitrogen.

In this stage the heat loss of the sun will be almost imperceptible. Owing to the low thermal conductive power of the earth's crust, there escapes through each square mile of this crust scarcely one-thousand-millionth part of the heat which the sun is radiating from an equal area of its surface. In future days, when the solar crust will have attained a thickness of 60 km. (40 miles), its loss of heat will be diminished to the same degree. The temperature on the surface of the sun may then still be some $50°$ or $60°$ above absolute zero, and volcanic eruptions will raise the temperature only for short periods and over small areas. Yet in the interior the temperature will still be at nearly the same actual intensity, something like several million degrees, and the compounds of infinite explosive energy will be stored up there as to-day. Like an immense dynamite magazine, the dark sun will float about in universal space without wasting much of its energy in the course of billions of years. Immutable, like a spore, it will retain its immense store of force until it is awakened by external forces into a new span of life similar to the old life. A slow shrinkage of the surface, due to the progressive loss of heat of the core and to the consequent contraction, will in the meanwhile have covered the sun with the wrinkles of old age.

Let us suppose that the crusts of the sun and the earth have the same thermal conductivity—namely, that of granite. According to Homén, a slab of granite one centimetre in thickness, whose two surfaces are at a temperature difference of 1° Cent., will permit 0.582 calorie to pass per minute per square centimetre of surface. By analogy, the earth's crust, with an increase of temperature of 30° per kilometre, as we penetrate inward, would allow $1.75 \cdot 10^{-4}$ calorie to pass per minute and per square centimetre (this is $\frac{1}{3580}$ of the mean heat supply of the earth, 0.625 calorie per minute per square centimetre); while the sun, with a crust of the same thickness as the earth, but with a diameter 108.6 times larger, would lose 3.3 times more heat per minute than the earth receives from it at the present time. At present the sun loses 2260 million times more heat than the earth receives; consequently, the loss of heat would be reduced to $\frac{1}{686,000,000}$ of the present amount. If the thickness of the solar crust amounted to $\frac{1}{140}$ of the solar radius—that is to say, to the same fraction that the thickness of the earth's crust represents of the terrestrial radius—the sun would in 74,500 million years not lose any more heat than it does now in a single year. This number has to be diminished, on account of the colder surface which the sun would have by that time, to about 60,000 million years. Considering that the mean temperature of the sun may be as high as 5 million degrees Celsius, the cooling down to the freezing-point of water might occupy 150,000 billion years, assuming that its mean specific heat is as great as that of water. During this time the crust of the sun would increase in thickness and the cooling would, of course, proceed at a decreasing rate. In any case, the total loss of energy during a period of a thousand billion years could, under these circumstances, only constitute a very small fraction of the total stored energy.

When an extinct star moves forward through infinite spaces of time, it will ultimately meet another luminous or likewise extinct star. The probability of such a collision is proportional to the angle under which the star appears—which, though very small, is not of zero magnitude —and to the velocity of the sun. The probability is increased by the deflection which these celestial bodies will undergo in their orbits on approaching each other. Our nearest neighbors in the stellar universe are so far removed from us that light, the light of our sun, requires, on an average, perhaps ten years to reach them. In order that the sun, with its actual dimensions and its actual velocity in space —20 km. (13 miles) per second—should collide with another star of similar kind, we should require something like a hundred thousand billion years. Suppose that there are a hundred times more extinct than luminous stars—an assumption which is not unjustifiable—the probable interval up to the next collision may be something like a thousand billion years. The time during which the sun would be luminous would represent perhaps one-hundredth of this—that is to say, ten billion years. This conclusion does not look unreasonable. For life has only been existing on the earth for about a thousand million years, and this age represents only a small fraction of the time during which the sun has emitted light and will continue to emit light. The probability of a collision between the sun and a nebula is, of course, much greater; for the nebulæ extend over very large spaces. In such a case, however, we need not apprehend any more serious consequences than result when a comet is passing through the corona of the sun. Owing to the very small amount of matter in the corona, we have not perceived any noteworthy effects in these instances. Nevertheless, the entrance of the sun into a nebula would increase the chance of a collision with another sun; for we shall see below that dark and luminous celestial bodies appear to be aggregated in the nebulæ.

From time to time we see new stars suddenly flash up in the sky, rapidly decrease in splendor again, to become extinguished or, at any rate, to dwindle down to faint visibility once more. The most remarkable of these exceedingly interesting events occurred in February, 1901, when a star of the first magnitude appeared in the constellation of Perseus. This star was discovered by Anderson, a Scotchman, on the morning of February 22, 1901. It was then a star of the third magnitude.[1] On a photograph which had been taken only twenty-eight

---

[1] Stars are classified in magnitude, the order being such that the most luminous stars have the lowest numbers. A star of the first magnitude is 2.52 times brighter than a star of the second magnitude; this, again, 2.52 times brighter than a star of the third magnitude, and so on. All this from the point of view of an observer on the earth.

hours previous to the discovery of this star, the star was not visible at all, although the plate marked stars down to the twelfth magnitude. The light intensity of this new star would hence appear to have increased more than five-thousand-fold within that short space of time. On February 23d the new star surpassed all other stars except Sirius in intensity. By February 25th it was of the first magnitude, by February 27th of the second, by March 6th of the third, and by March 18th of the fourth magnitude. Then its brightness began to fluctuate periodically up to June 22d, with a period first of three, then of five days, while the average light intensity decreased. By June 23d it was of the sixth magnitude. The light intensity diminished then more uniformly. By October, 1901, it was a star of the seventh magnitude; by February, 1901, of the eighth magnitude; by July, 1902, of the ninth magnitude; by December, 1902, of the tenth magnitude; and since then it has gradually dwindled to the twelfth magnitude. When this star was at its highest intensity it shone with a bluish-white light. The shade then changed into yellow, and by the beginning of March, 1901, into reddish. During its periodical fluctuations the hue was whitish yellow at its maximum and reddish at its minimum intensity. Since then the color has gradually passed into pure white.

The spectrum of this star shows the greatest similarity to that of the new star in the constellation Auriga (Nova Aurigæ) of the year 1892 (Fig. 45).

On the whole, it is characteristic of new stars that the spectrum lines appear double—dark on the violet and bright on the red side. In the spectrum of Nova Aurigæ this peculiarity is, among others, striking in the three hydrogen lines C, F, and H, in the sodium line, in the nebula lines, and also in the magnesium line. In the spectrum

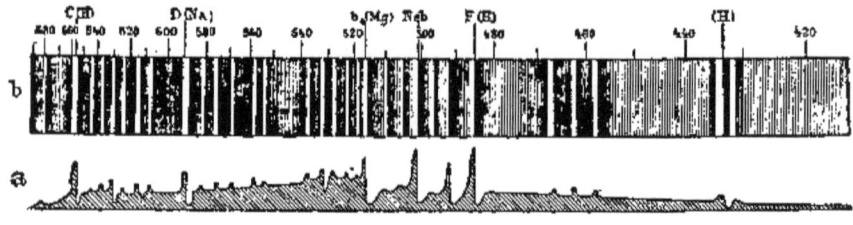

Fig. 45.—Spectrum of Nova Aurigæ, 1892

of Nova Persei the displacement of the hydrogen lines towards the violet is so great that, according to Doppler's principle,[1] the hydrogen gas which absorbed the light must have been moving towards us with a velocity of 700 or more kilometres (450 miles) per second. Some calcium lines show a similar displacement, which is less noticeable in the case of the other metals. This would

[1] When, standing on a station platform, we watch an express train rushing through the station, the pitch of the engine whistle seems to become higher as long as the train is approaching us, and deeper again when the train is moving away from us. The pitch of a note depends upon the number of oscillations which our ear receives per second. Now, when the train is fast approaching us, more vibrations are sent into our ear than when the train is at a stand-still, and the pitch, therefore, appears to become higher. The same reasoning holds for light waves, of which Doppler, of Prague (Bohemia), was in fact thinking when first announcing his principle in 1842. The wave-length of a particular color of the spectrum is fixed with the aid of some Fraunhofer line characteristic of a certain metal. If we compare the spectrum of a star and the spectrum of a glowing metal, photographed on the same plate, the stellar lines will appear shifted towards the violet end (violet light is produced by nearly twice as many vibrations of the ether per second as red light) when the star is moving towards us in the line of sight. This principle has successfully been applied by Huggins, H. C. Vogel, and others, for determining the motion of a star in our line of sight. When a star is revolving about its own axis, the equatorial belt will seem to come nearer to us (or to recede from us), while the polar regions will seem to be at a stand-still; the lines will then appear oblique (not vertical). In this way Keeler proved that the rings of Saturn consists of swarms of meteorites moving at different velocities in the different rings.—H. B.

appear to indicate that relatively cold masses of gas are issuing from the stars and streaming with enormous velocities towards the earth. The luminous parts of the stars were either at a stand-still or they were moving away from us. The simplest explanation of these phenomena would be that a star when flashing up by virtue of its high temperature and high pressure shows enhanced (widened) spectral lines, whose violet portion is absorbed by the strongly cooled masses of gas which are moving towards us and are cooled by their own strong expansion. These gases stream, of course, in all directions from the star, but we only become aware of those gases which absorb the light of the stars—that is to say, those situated between the star and the earth, and streaming in our direction.

Gradually the light of the metallic lines and of the continuous spectrum on which they were superposed began to fade, first in the violet, while the hydrogen lines and nebular lines still remained distinct; like other new stars, this star showed, after a while, the nebular spectrum. This interesting fact was first noticed by H. C. Vogel in the new star in the Swan (Nova Cygni, 1876). The star P in the Swan, which flashed up in the year 1600, still offers us a spectrum which indicates the emission of hydrogen gas. It is not impossible that this "new" star has not yet reached its equilibrium, and is still continuing to emit cold streams of gases. Insignificant quantities of gas suffice for the formation of an absorption spectrum; thus the emission of gas might continue for long periods without exhausting the supply.

We have already mentioned (page 116) the peculiar clouds of light which were observed around Nova Persei. Two annular clouds moved away from this star with velocities of 1.4 and 2.8 seconds of arc per day between March

29, 1901, and February, 1902. If we calculate backward from these dates the time which must have elapsed since those gases left the star, we arrive at the date of the week—February 8 to 16, 1901—in satisfactory agreement with the period of greatest luminosity of the star of February 23d. It would, therefore, appear that these emanations came from the star and were ejected by the radiation pressure. Their light did not mark any noticeable polarization, and could not be reflected light for this reason. We may suppose that the dust particles discharged their electric charges, and that the gases became thereby luminous.

In this case we were evidently witnesses of the grand *finale* of the independent existence of a celestial body by collision with some other body of equal kind. The two colliding bodies were both dark, or they emitted so little light that their combined light intensities did not equal that of a star of the twelfth magnitude. As, now, their splendor after the collision was greater than that of a star of the first magnitude, although their distance has been estimated to be at least 120 light years,[1] their radiation intensity must have exceeded that of the sun several thousand times. Under these circumstances the mechanical radiation pressure must also have been many times more powerful than on the surface of the sun, and the masses of dust which were ejected by the new star must have possessed a velocity very much greater than that of solar dust. Yet this velocity must have been smaller than that of light, which, indeed, the effect of the radiation pressure can never equal.

It is not difficult to picture to ourselves the enormous violence with which this "collision" must have taken

[1] One light year corresponds to 9.5 billions kilometres, and it is the distance which the light traverses in the course of a year.

ORIGIN OF NEBULÆ

place. A strange body — for instance, a meteorite — which rushes from the infinite universe into the sun has at its collision a velocity of 600 km. (400 miles) per second, and the velocity of the two colliding suns must have been of approximately that order. The impact will in general be oblique, and, although part of the energy will of course be transformed into heat, the rest

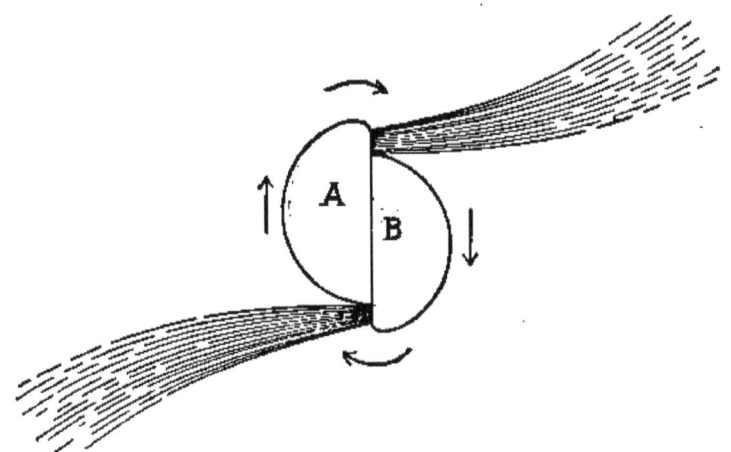

Fig. 46.—Diagram indicating the consequences of a collision between two extinct suns, A and B 'moving' in the direction of the straight arrows. A rapid rotation in the direction of the curved arrows results, and two powerful streamers are ejected by A B, the explosive substances from the deeper strata of A and B being brought up to the surface by the collision

of the kinetic energy must have produced a rotational velocity of hundreds of kilometres per second. By comparison with this number the actual circumferential speed of the sun, about 2 km. (1¼ miles) per second on the equator, would vanish altogether; and the difference is still more striking for the earth, with its 0.465 km. per second at the equator. We shall, therefore, not commit an error of any consequence if we presume the two bodies to have been practically devoid of circumferential speeds before their collision. At the collision, matter will have been ejected from both these celestial bodies at right angles to the relative directions of their motion in two powerful torrents, which would be situated in the plane in which the two bodies were approaching each other (compare Fig. 46). The rotational speed of the double star, which will be diminished by this ejection of matter, will have contributed to increase the energy of ejection. We remember, now, that when matter is brought up from the interior to the surface of the sun it will behave like an explosive of enormous power. The ejected gases will be hurled in terrific flight about the rapidly revolving central portions. We obtain an idea (though a very imperfect one) of these features when we look at a revolving pinwheel in a fireworks display. Two pinwheels have been attached to the ends of a diameter and belch forth fire in radial lines. The farther removed from the wheel, the smaller will be the actual velocity and also the angular velocity of these torrents of fire. Similarly with the star. The streams are rapidly cooled, owing to the rapid expansion of the gas. They will also contain fine dust, largely consisting of carbon, probably, which had been bound by the explosive materials. The clouds of fine dust will obscure the new star more and more, and will gradually change its white brilliancy into yellow and reddish, because the fine dust weakens blue-and-green rays more than it does yellow-and-red rays. At first the clouds were so near to the the star that they possessed a high angular velocity of their own; they then appeared to surround the star completely. But after March 22, 1901, the outer particles of the streams attained greater distances and assumed longer periods of revolution (six days); the star then became more obscured when the

ORIGIN OF NEBULÆ

extreme dust clouds of the streams covering it happened to get between us and the star. As the streams of particles were moving farther away, their rotational periods increased gradually to ten days. The star, therefore, became periodical with a slowly growing length of period, and its glow turned more reddish at its minimum than at its maximum of intensity. At the same time, the absorptive power of the marginal particles decreased, partly owing to their increasing expansion, partly because the dust was slowly aggregating to coarser

Fig. 47.—Spiral nebula in the Canes Venatici. Messier 51. Taken at the Yerkes Observatory on June 3, 1902. Scale, 1 mm. = 13.2 sec. of arc

particles; possibly, also, because the finest particles were being driven away by the radiation pressure. The sifting influence which the dust exercised upon the light, and owing to which the red-and-yellow rays were more readily transmitted than the blue-and-green, gradually became impaired; hence the color of the light turned more gray, and after a certain time the star appeared once more of a whitish hue. This white color indicates that the star must still have a very high temperature. By the continued ejection of dust-charged masses of gas, probably with gradually decreasing violence, the light intensity of the star must slowly diminish (as seen from the earth) and the distribution of the layers of dust around the luminous core will more and more become uniform. How violent the explosion must have been, we recognize from the observation that the first ejected masses of hydrogen rushed out with an apparent velocity of at least 700 km. per second. This velocity is of the same order as that of the most remarkable prominences of the sun.

It will be admitted that these arguments present us with a faithful simile even of the details of the observed course of events, and it is therefore highly probable that our view is in the main correct. But what has meanwhile become of the new star? Spectrum analysis tells us that it has been converted into a stellar nebula like other new stars. The continuous light of the central body has more and more been weakened by the surrounding masses of dust. By the radiation pressure these masses are driven towards the outer particles of the surrounding gaseous envelope consisting principally of hydrogen, helium, and "nebular matter." There the dust discharges its negative electricity, and thus calls forth a luminescence which equals that of the nebulæ.

Fig. 48.—Spiral nebula in the Triangle. Messier 33. Taken at the Yerkes Observatory on September 4 and 6, 1902. Scale, 1 mm.= 30.7 sec. of arc

We have to consider next that owing to the incredibly rapid rotation, the central main mass of the two stars will, in its outer portions, be exposed to centrifugal forces of extraordinary intensity, and will therefore become flattened out to a large revolving disk.[1] As the pressure in the outer portions will be relatively small, the density of the gases will likewise be diminished there. The energetic expansion and, more still, the great heat radiation will lower the temperature at a rapid rate. We have thus to deal with a central body whose inner portion will possess a high density, and which will resemble the mass of the sun, while the outer portion will be attenuated and nebular. Distributed about the central body we shall find the rest of the two streams of gases which were ejected immediately after the violent collision between the two celestial bodies. A not inconsiderable portion of the matter of these spirally arranged outer parts will probably travel farther away into infinite space, finally to join some other celestial body or to form parts of the great irregular spots of nebular matter which are collected around the star clusters. Another portion, not able to leave the central body, will remain in circular movement about it. In consequence of this circular movement, which will be extremely slow, the outlines of the two spirals will gradually become obliterated, and the spirals will themselves more and more assume the shape of nebular rings about the central mass.

This spiral form (Figs. 47 and 48) of the outer portions

---

[1] A. Ritter has calculated that when two suns of equal size collide with one another from an infinite distance, the energy of the collision is not more than sufficient to enlarge the volume of the suns to four times the previous amount. The largest portion of the mass will therefore probably remain in the centre, and it will only be masses of light gases which will be ejected.

Fig. 49.—The great nebula in Andromeda. Taken at the Yerkes Observatory on September 18, 1901. Scale, 1 mm. = 54.6 sec. of arc

of the nebulæ has for a long time excited the greatest attention. In almost all the investigated instances it has been observed that two spiral branches are coiling about the central body. This would indicate that the matter is in a revolving movement about the central axis of the spiral, and that it has streamed away from the axis in two opposite directions. Sometimes the matter appears arranged as in a coil; of this type the great nebula of Andromeda is the best-known example (Fig. 49). A closer inspection of this nebula with more powerful instruments indicates, however, that it is also spiral and that it appears coiled, because we are looking at it from

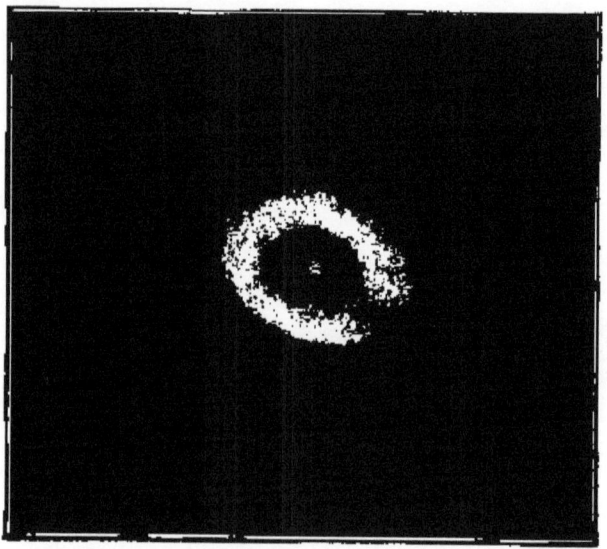

Fig. 50.—Ring-shaped nebula in Lyra. Taken at the Yerkes Observatory.

the side. The late famous American astronomer Keeler, who has studied these nebulæ with greater success than any one else, has catalogued a great many of them in all the divisions of the heavens which were accessible to his

ORIGIN OF NEBULÆ

instruments, and he has found that these formations are predominatingly of a spiral nature.

Some nebulæ, like the so-called planetary nebulæ, offer rather the appearance of luminous spheres. We

Fig. 51.—Central portion of the great nebula in Orion. Taken at the Yerkes Observatory. Scale, 1 mm.=12 sec. of arc

may assume in these cases that the explosions were less violent, and that the spirals, therefore, are situated so closely together that they seem to merge into one another. Possibly the inequalities in their development have become equalized in the course of time. A few nebulæ are ring-shaped, as the well-known nebula of Lyra (see Fig. 50). These rings may, again, have been formed out of spiral nebulæ, and the spirals may have gradually been obliterated by rotation, while the central nebulous matter may have been concentrated on the planets travelling round the central star. Schaeberle, an eminent American astronomer, has discovered traces of spiral shape also in the Lyra nebula.

Another kind of nebula is the ordinary nebula of vast extension and irregular shape, evidently formed out of most extremely attenuated matter; well-known characteristic examples are found in Orion, about the Pleiades, and in the Swan (Figs. 51, 52, and 53). In these nebulæ portions of a spiral structure have likewise often been discerned.

We have said that the collision between two celestial bodies would result in the formation of a spiral with two wings. If the impact is such that the two centres of the celestial bodies move straight towards each other, a disk will arise, and not a spiral; or if one star is much smaller than the other, possibly a cone, because the gases will uniformly be spread in all directions about the line of impact. A perfectly central impact is obviously very rare; but there may be cases which approach this limiting condition more or less, especially when the relative velocity of the two bodies is small. By slow diffusion a feebly developed spiral may also be converted into a rotating disklike structure. The extension of these nebular structures will depend upon the ratio between the mass of the system and the velocity of ejection of the gases. If, for example, two extinct suns of nearly equal dimensions and mass, like our sun, should collide, some gas masses would travel into infinite space, being hurled out with a velocity of more than 900 km. (550 miles) per second; while other particles, moving at a slower rate, would remain in the neighborhood of the

Fig. 52.—Nebular striæ in the stars of the Pleiades. Taken at the Yerkes Observatory on October 19, 1901. Scale, 1 mm.=42.2 sec. of arc

central body. The nearer to that body, the smaller was their velocity. From their position they might fall back into the central body, to be reincorporated in it, if two circumstances did not prevent this. The one circumstance is the enormous radiation pressure of the glowing central mass. That pressure keeps masses of dust particles floating, which by friction will carry the surrounding masses of gas with them. Owing to the absorption of the radiation by the dust particles, only the finer particles will be supported farther outside, and at the extreme margin of the nebula even the very finest dust will no longer be maintained in suspension by the greatly weakened radiation pressure. Thus we arrive at an outer limit for the nebula. The other circumstance is the violent rotation which is set up by the impact of the central bodies. The rotation and the centrifugal forces will produce a disk-shaped expansion of the whole central mass. Owing to molecular collisions and to tidal effects, the angular velocity will in the denser portions tend to become uniform, so that the whole will rotate like a flattened-out ball filled with gas, and the spiral structure will gradually disappear in those parts. In the more remote particles the velocity will only increase to such an extent as to equal that of a planet moving at the same distance—that is to say, the gravitation towards the central body will be balanced by the centrifugal force, and at the very greatest distances the molecular bombardments, as well as gravitation towards the centre, will become so insignificant that any masses collected there will retain their shape for an almost unlimited space of time.

In the centre of this system the main bulk of the matter would be concentrated in a sun of extreme brightness, whose light intensity would, however, owing

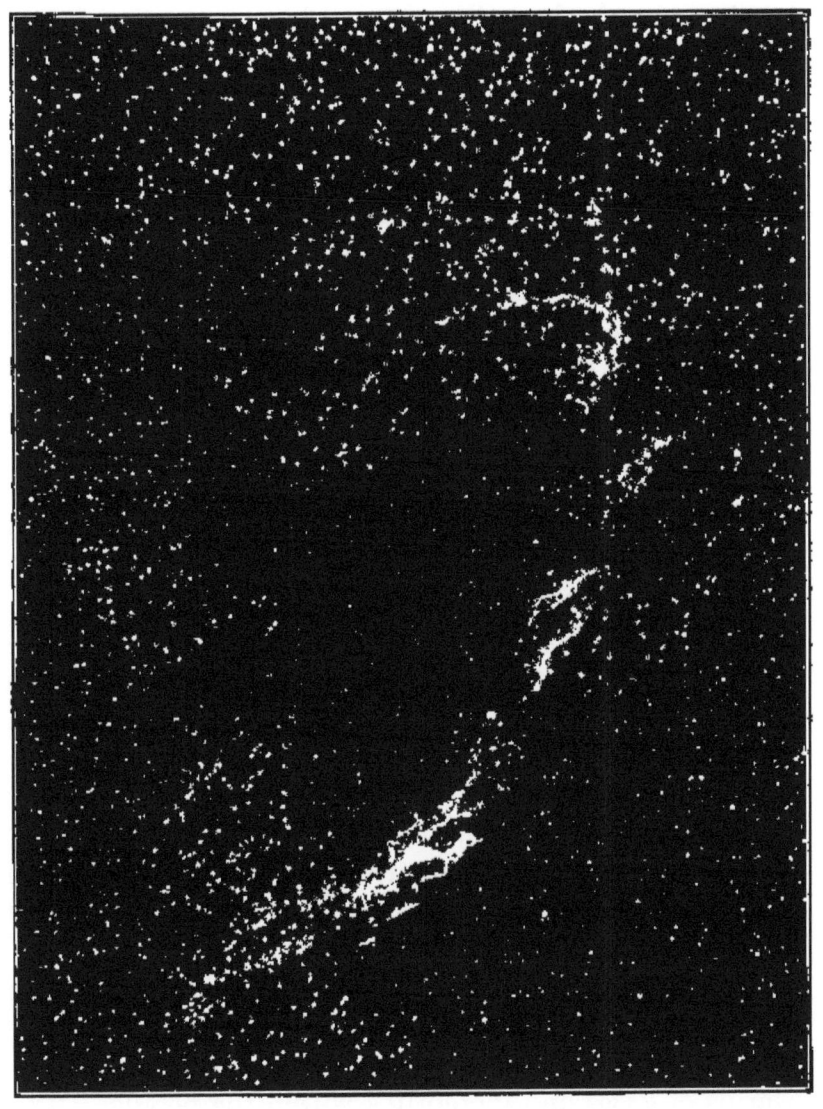

Fig. 53.—Nebular striæ in the Swan. New General Catalogue, 6992. Taken at the Yerkes Observatory on October 5, 1901. Scale, 1 mm. =41 sec. of arc

to strong radiation, diminish with comparative rapidity.

Such an extensive nebular system, in which gravitation, on account of the enormous distances, would act feebly and very slowly, would yet, in spite of the extraordinary attenuation of matter in its outer portions, and just on account of its vast extension, be able to stop the movement of the particles of dust penetrating into it. If the gases of the nebula are not to escape into space, notwithstanding the infinitesimal gravitation, their molecules must be assumed to be almost at a stand-still, and their temperature must not rise by more than 50° or 60° Cent. above absolute zero. At such low temperatures the so-called adsorption plays an enormously important part (Dewar). The small dust particles form centres about which the gases are condensed to a remarkable degree. The extremely low density of these gases does not prevent their condensation; for the adsorption phenomenon follows a law according to which the mass of condensed gas will only be reduced by about one-tenth when the density of the surrounding gas has been decreased by one-ten-thousandth. The mass of dust particles or dust grains will thus be augmented, and when they collide they will be cemented together by the semi-liquid films condensed upon them. There must, hence, be a relatively energetic formation of meteorites in the nebulæ, and especially in their interiors. Then stars and their satellites, migrating through space, will stray into these swarms of gases and meteorites within the nebulæ. The larger and more rapidly moving celestial bodies will crush through this relatively less dense matter; but thousands of years may yet be occupied in their passing through nebulæ of vast dimensions.

An extraordinarily interesting photograph obtained by

Fig. 54.—Nebula and star rift in the Swan, in the Milky Way. Taken by M. Wolf, Königstuhl, near Heidelberg

WORLDS IN THE MAKING

the celebrated Professor Max Wolf, of Heidelberg, shows us a part of the nebula in the Swan into which a star has penetrated from outside. The intruder has collected about it the nebulous matter it met on its way, and has thus left an empty channel behind it marking its track. Similar spots of vast extent, relatively devoid of nebulous matter, occur very frequently in the irregular nebulæ; they are frequently called "fissures," or by the specifi-

Fig. 55.—Great nebula near Rho, in Ophiuchus. Photograph by E. E. Barnard, Lick Observatory There are several empty spots and rifts near the larger stars of the nebula

ORIGIN OF NEBULÆ

cally English term "rifts," because they have generally a long-drawn-out appearance. The presumption that these rifts represent the tracks of large celestial bodies

Fig. 56. — Star cluster in Hercules. Messier 13. Taken at the Yerkes Observatory. Scale, 1 mm.=9.22 sec. of arc

which have cut their way through widely expanded nebular masses (Fig. 54) has been entertained for a long time.

The smaller and more slowly moving immigrants, on the other hand, are stopped by the particles of the nebulæ. We therefore see the stars more sparsely distributed in the immediate neighborhood of the nebulæ, while in the nebulæ themselves they appear more densely crowded. This fact had struck Herschel in his observations of nebulæ; in recent days it has been investigated by Courvoisier and M. Wolf. In this way several centres of attraction are created in a nebula; they condense the gases surrounding the nebula, and catch, so to say, any stray meteorites and collect them especially in the inner portions of the nebula. We frequently observe, further, how the nebular matter appears attenuated at a certain distance from the luminous bright stars (compare Figs. 52 and 55). Finally, the nebulæ change into star clusters which still retain the characteristic shapes of the nebulæ; of these the spiral is the most usual, while we also meet with conical shapes, originating from conical nebulæ, and spherical shapes (compare Figs. 56, 57, and 58).

This is, broadly, the type of evolution through which Herschel, relying upon his observations, presumed a nebula to pass. He was, however, under the impression that the nebulous matter would directly be condensed into star clusters without the aid of strange celestial immigrants.

It has been known since the most ancient times, and has been confirmed by the observations of Herschel and others in a most convincing manner, that the stars are strongly concentrated about the middle line of the Milky Way. It is not improbable that there was originally a nebula of enormous dimensions in the plane of the Milky Way, produced possibly by the collision of two such giant suns as Arcturus. This gigantic nebula has gathered up the smaller migrating celestial bodies which, in their turn, have condensed upon themselves nebular matter, and have thereby become incandescent, if they were not so before. The rotational movement in those parts which were far removed from the centre of the Milky Way may be neglected. At a later period collisions succeeded between the single stars which had been gathered up, and it is for this reason that gaseous nebulæ, as well as new stars, are comparatively frequent phenomena in the plane of the Milky Way. This view may some day receive

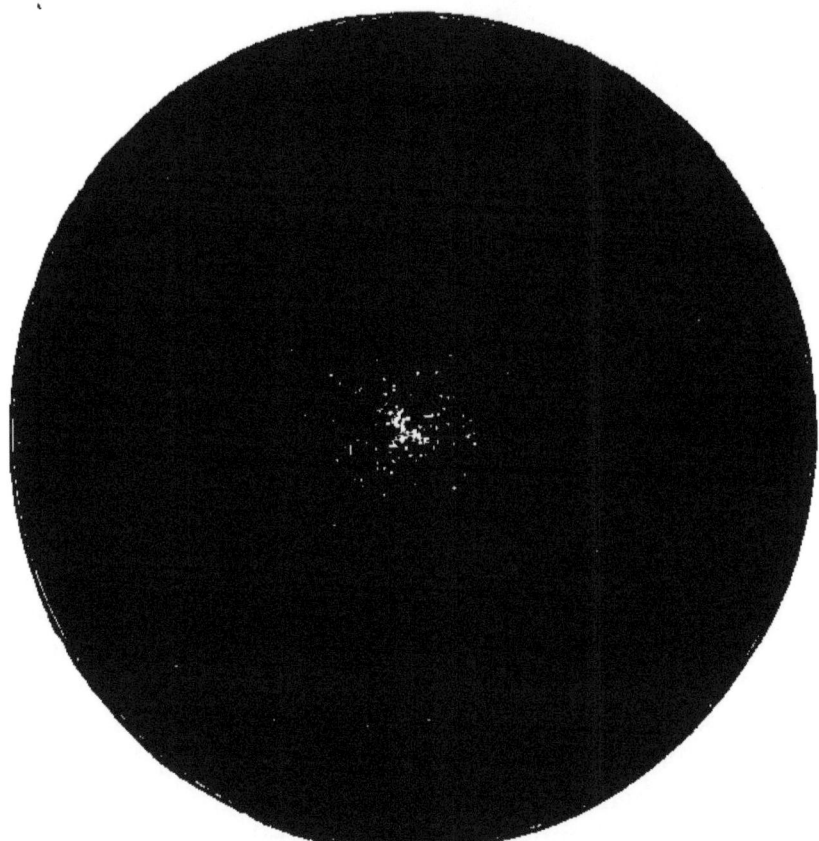

Fig. 57.—Star cluster in Pegasus. Messier 15. Taken at the Yerkes Observatory. Scale, 1 mm.=6.4 sec. of arc

confirmation, when we succeed in proving the existence of a central body in the Milky Way, evidence of which might possibly be deduced from the curvature of the orbits of the sun or of other stars.

As regards the ring-shaped nebula in the Lyre (Fig. 50), the most recent measurements made by Newkirk point to the result that the star visible in its centre is distant from us about thirty-two light-years. As it appears probable that this star really forms the central core of the nebula, the distance of the nebula itself must be thirty-two light-years. From the diameter of the ring-shaped nebula which Newkirk estimates at one minute of arc, this astronomer has calculated that the distance of the ring from its central body is equal to about three hundred times the radius of the earth's orbit—that is to say, the ring is about ten times as far from its sun as Neptune is from our sun. There is a faint luminescence within this ring. The nebular matter may originally have been more concentrated at this spot than in the outer portions of the ring itself. But this mass was probably condensed on meteors which immigrated from outside, and when these meteors coalesced dark planets were produced which move about the central body, and which have gathered about them most of the gases. If that central body were as heavy as our sun, the matter in the ring should revolve about it in five thousand years. That rotation would suffice to wipe out the original spiral shape, enough of which has yet been left to permit of our

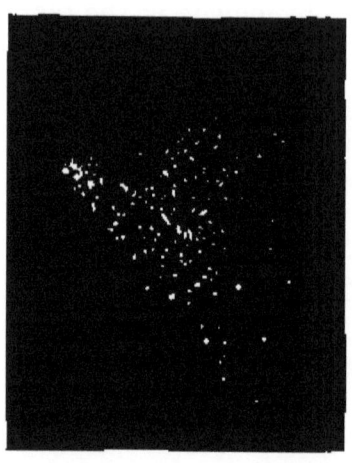

Fig. 58.—Cone-shaped star cluster in Gemini.

ORIGIN OF NEBULÆ

distinctly discerning the two wings of the spiral. The central body of this ring-shaped nebula gives a continuous spectrum of bright lines which is particularly developed on the violet side. The star would therefore appear to be much younger and much hotter than our sun, and its radiation pressure would therefore be much more intense. The period of rotation of the nebula may, for this reason, have to be estimated at a considerably higher figure.

The eminent Dutch astronomer Kapteyn has deduced from the proper motions of 168 nebulæ that their average distance from the earth is about seven hundred light-years and equal to that of stars of the tenth magnitude. The old idea, that the nebulæ must be infinitely farther removed from us than the fainter stars, would therefore appear to be erroneous. According to the measurements of Professor Bohlin, the nebula in Andromeda may indeed be at a distance of not more than forty light-years.

The "new stars" form a group among the peculiar celestial bodies which on account of their variable light intensity have been designated as "variable stars," and of which a few typical cases should be mentioned, because a great scientific interest attaches to these problems. The star Eta, in Argus, may be said to illustrate the strange fate that a star has to pass through when it has drifted into a nebula filled with immigrated celestial bodies. It is one of the most peculiar variable stars. The star shines through one of the largest nebular clouds in the heavens. Whether it stands in any physical connection with its surroundings cannot be stated without further examination. The star might, for instance, be at a considerable distance in front of the nebula, between the latter and ourselves. Its frequent change in light intensity suggests, however, a series of collisions, which do not appear unnatural to us when we suppose that the star is within a nebula into which many celestial bodies have drifted.

As this star belongs to the southern hemisphere, it was not observed before our astronomers commenced to visit that hemisphere. In 1677 it was classed as a star of the fourth magnitude; ten years later it was of the second magnitude; the same in 1751. In 1827 it was of the first magnitude, and it was found to be variable—that is to say, it shone with variable brightness. Herschel observed that it fluctuated between the first and second magnitudes, and that it increased in brightness after 1837, so that it was by 1838 of magnitude 0.2. After that it began to decrease in intensity up to April, 1839, when it had the magnitude 1.1. It remained for four years approximately at this intensity; then it increased rapidly again in 1843, and surpassed all stars except Sirius (magnitude -1.7).[1] Afterwards its intensity slowly diminished once more, so that it remained just visible to the naked eye (sixth magnitude); by 1869 it had become invisible. Since then it has been fluctuating between the sixth and seventh magnitudes.

The last changes in the intensity of this star strongly recall the behavior of the new star in Perseus, only that the latter has been passing through its phases at a much more rapid rate. It appears to be certain, however, that Eta, in Argus, was from the very beginning far brighter than Nova Persei, and that at least once before the great collision in 1843 (after which it was surrounded by obscuring clouds of increasing opacity)—namely, in January, 1838, it had been exposed to a slight collision of quickly

---

[1] This figure, -1.7, signifies that the brightness of Sirius is $2.52^{2.7} = 12$ times greater than that of a star of magnitude 1. Next to Sirius comes Canopus, with magnitude -1.0, being 6.3 times brighter than a star of magnitude 1.

vanishing effect. This lesser collision was probably of the kind which Mayer imagined for the earth and sun. It would give rise to heat development corresponding to the heat expenditure of the sun in about a hundred years. As it had been observed that the star was variable in an irregular manner before that, we may, perhaps, presume that it had already undergone another collision.

According to the observations of Borisiak, a student in Kief, the new star in Perseus would have been, on the evening of February 21, 1901, of 1.5 magnitude, while a few hours previously it had been of magnitude 12, and the following evening of magnitude 2.7; afterwards its intensity increased up to the following evening, when it outshone all the other stars in the northern sky. If this statement is not based on erroneous observations, the new star must have been in collision with another celestial body two days before its great collision, perhaps with a small planet in the neighborhood of the sun, with which it later collided. That would account for its temporary brilliancy.

New stars are by no means so rare as one might perhaps assume. Almost every year some new star is discovered. By far most of these are seen in the neighborhood of the Milky Way, where the visible stars are unusually crowded, so that a collision which would become visible to us may easily occur.

For similar reasons we find there also most of the gaseous nebulæ.

Most of the star clusters are also in the neighborhood of the Milky Way. This is in consequence of the facts just alluded to. The nebulæ which are produced by collisions between two suns are soon crossed by migrating celestial bodies such as meteorites or comets, which there occur in large numbers; by the condensing action of these intruders they are then transformed into star clusters. In parts of the heavens where stars are relatively sparse (as at a great distance from the Milky Way), most of the nebulæ observed exhibit stellar spectra. They are nothing but star clusters, so far removed from us that the separate stars can no longer be distinguished. That single stars and gaseous nebulæ are so rarely perceived in these regions is, no doubt, due to their great distance.

Among the variable stars we find quite a number which display considerable irregularity in their fluctuations of brightness, and which remind us of the new stars. To this class belongs the just-mentioned star Eta, in Argus. Another example (the first one which was recognized as "variable") is Mira Ceti, which may be translated, "The Wonderful Star in the Constellation of the Whale." This mysterious body was discovered by the Frisian priest Fabricius, on August 12, 1596, as a star of the second magnitude. The priest, an experienced astronomer, had not previously noticed this star, and he looked for it in vain in October, 1597. In the years 1638 and 1639 the variability of the star was recognized, and it was soon ascertained to be irregular. The period has a length of about eleven months, but it fluctuates irregularly about this figure as a mean value. At its greatest intensity the star ranks with those of the first or second order. Sometimes it is weaker, but it is always of more than the fifth magnitude. Ten weeks after a maximum the star is no longer visible, and its brightness may diminish to magnitude 9.5. In other words, its intensity varies about in the ratio of 1 : 1000 (or possibly more). After a minimum the brightness once more increases, the star becomes visible again—that is to say, it attains the sixth magnitude — and after another six weeks it will once

ORIGIN OF NEBULÆ

more be at its maximum. We have evidently to deal with several superposed periods.

The spectrum of this star is rather peculiar. It belongs to the red stars with a band spectrum which is crossed by bright hydrogen lines. The star is receding from us with a velocity of not less than 63 km. (39 miles) per second. The bright hydrogen lines which correspond to the spectrum of the nebula may sometimes be resolved into three components, of which the middle one corresponds to a mean velocity of 60 km., and the two others have variable receding velocities of 35 and 82 km.—that is to say, velocities of 25 or 22 km. less or more than the mean velocity. Evidently the star is surrounded by three nebulæ; one is concentrated about its centre; the two others lie on a ring the matter of which has been concentrated on two opposite sides. The ring, which recalls the ring nebula in the Lyre, seems to move about the star with a velocity of 23.5 km. per second. As this revolution is accomplished within eleven—or, more correctly, within twenty-two months, since there must be two maxima and two minima during one revolution—the total circumference of the ring will be 23.5 × 86,400 × 670—1361 millions, and the radius of its orbit 217 million km., which is 1.45 times greater than the radius of the earth's orbit. Now the velocity of the earth in its orbit is 29.5 km. (18.3 miles) per second. A planet at 1.45 times that distance from the sun would have the (1.203 times smaller) velocity of 24.5 km. per second, which is approximately that of the hypothetical ring of Mira Ceti. We conclude, therefore, that the mass of the central sun in Mira Ceti will nearly equal the mass of our sun. The calculation really suggests that Mira would be about eight per cent. smaller; but the difference lies within the range of the probable error.

Chandler has directed attention to a striking regularity in these stars. The longer the period of their variation, the redder in general their color. This is easily comprehended. The denser the original atmosphere, the more widely the gases will have extended outward from the star, and the more dust will have been caught or secreted by it. We have seen that the limb of the sun has a reddish light because of the quantities of dust in the solar atmosphere. The effect is chiefly to be ascribed to the absorption of the blue rays by the dust; but it may partly be explained on the assumption that the solar radiations render the dust incandescent, though its temperature may be lower than that of the photosphere, because the dust lies outside the sun, and that it will therefore emit a relatively reddish light. The more dust there is in a nebula, the redder will be its luminescence; and as the quantity of dust increases in general with the extension of the nebula, that star which is surrounded by wider rings of nebulæ will in general be more red; but the greater the radius of the ring, the longer also will in general be its period.

The so-called red stars show, in addition to the bright hydrogen lines, banded spectra which indicate the presence of chemical compounds. On this account such stars were formerly credited with a lower temperature. But the same peculiarity is also observed in sun-spots, although the latter, on account of their position, must have a higher temperature than the surrounding photosphere. The presence of bands in the spectrum certainly suggests high pressure, however. The red stars are evidently surrounded by a very extensive atmosphere of gases, in the inner portions of which the pressure is so high that the atoms enter into combination. The spectra of the red stars display, on the whole, a striking resemblance to those of the sun-spots. The violet portion of the spectrum is weakened, because the masses of dust have extinguished this light. Owing to the large masses of dust which lie in our line of sight, the spectrum lines are in both cases markedly widened and sometimes accompanied by bright lines.

Another class of stars, distinguished by bright lines, comprises those studied by Wolf and Rayet, and named after them. These stars are characterized by a hydrogen atmosphere of enormous extension, large enough in some cases, it has been calculated, to fill up the orbit of Neptune. These stars are evidently either hotter and more strongly radiating than the red stars, or there is not so much dust in their neighborhood—the dust may possibly have been expelled by the strong radiating pressure. They are, therefore, classed with the yellow, and not with the red stars. Although there is every reason to suppose that their central bodies are at least as hot as those of the white stars, the dust is yet able to reduce the color to yellow, owing to the vast extensions of their atmospheres.

The unequal periods in stars like Mira may be explained by the supposition that there are several rings of dust moving about them, as in the case of the planet Saturn. In the case of the inner rings which have a short period, there has probably been sufficient time during the uncounted number of revolutions to effect a uniform distribution of the dust. Hence we do not discern any noteworthy nuclei in them, such as we observe in the tails of comets; the dust rings only help to impart to the star a uniform reddish hue. In the outer rings the distribution of dust will, however, not be uniform. One of the rings may be responsible for the chief proper period. By the co-operation of other less important dust rings, the maximum or minimum, we shall easily understand, may be displaced, and thus the time interval between the maxima and minima be altered. This alteration of the period is so strong for some stars that we have not yet succeeded in establishing any simple periodicity. The best-known star of this type is the bright-red star Betelgeuse in the constellation of Orion. The brightness of this star fluctuates irregularly between the magnitudes 1.0 and 1.4.

By far the largest number of variable stars belong to the type of Mira. Others resemble the variable star Beta in the constellation of the Lyre, and thus belong to the Lyre type. The variability of the spectra of a great many of these stars indicates that they must be moving about a dark star as companion, or rather that they both move about a common centre of gravity. The change in the light intensity is, as a rule, explained by the supposition that the bright star is partially obscured at times by its dark companion. Many irregularities, however, in their periods and other circumstances prove that this explanation is not sufficient. The assumption of rings of dust circulating about the star and of larger condensation centres affords a better elucidation of the variability of these stars. They are grouped with the white or yellow stars, in whose surroundings the dust does not play so large a part as in that of Mira Ceti. The period of their variability is, as a rule, very short, moreover—generally only a few days (the shortest known, only four hours)—while the period of the Mira stars amounts to at least sixty-five days, and may attain two years. There may be still longer periods so far not investigated.

Nearly related to the Lyre stars are the Algol stars, whose variability can be explained by the assumption that another bright or dark star is moving within their vicinity, partially cutting off their light. There is no dust in these cases, and the spectrum characterizes these stars as stars of the first class—that is, as white stars—so far as they have been studied up to the present.

We must presume for all the variable stars that the line of sight from the observer to the star falls in the plane of their dust rings or of their companions. If this were not so, they would appear to us like a nebula with a central condensation nucleus, or, so far as Algol stars are concerned, like the so-called spectroscopic doubles whose motion about each other is recognized from the displacement of their spectral lines.

The evolution of stars from the nebulous state has been depicted by the famous chief of the Lick Observatory, in California, W. W. Campbell, as follows (compare the spectra of the stars of the 2d, 3d, and 4th class, Figs. 59 and 60):

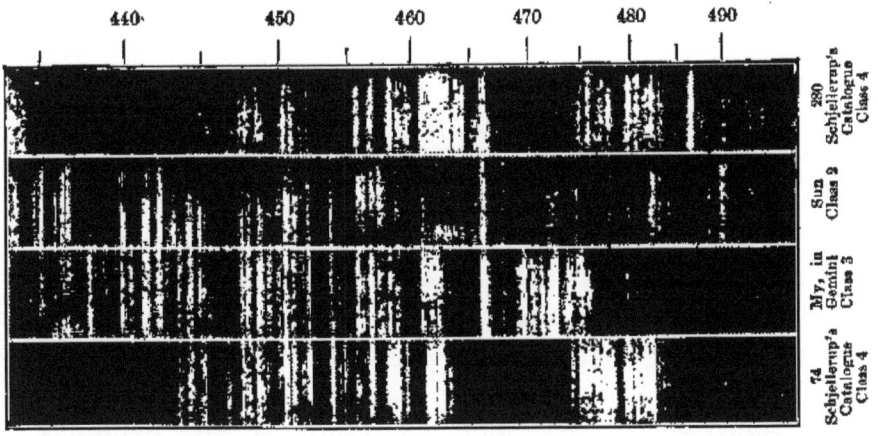

Fig. 59.—Comparison of spectra of stars of classes 2, 3, 4. After photographs taken at the Yerkes Observatory. Blue portions of spectrum. Wave-lengths in millionths of a millimetre

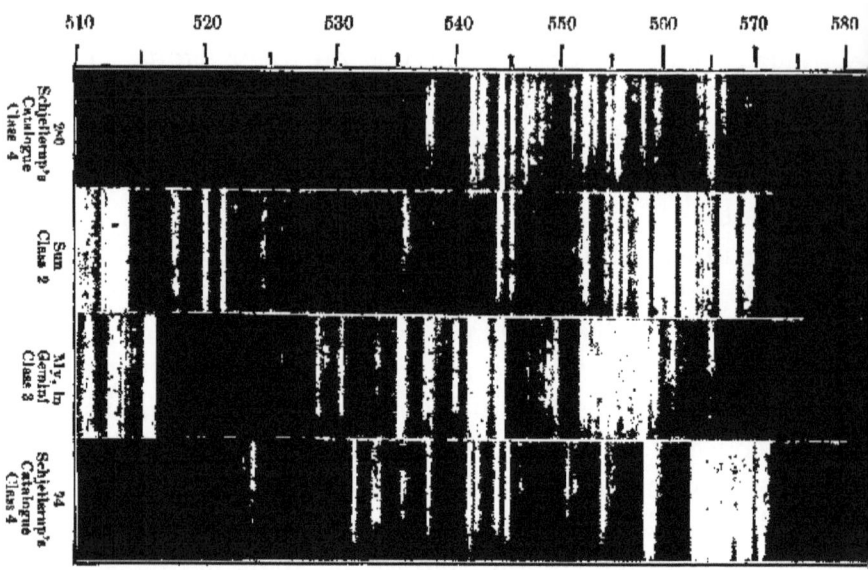

Fig. 60.—Comparison of spectra of stars of classes 2, 3, 4. After photographs taken at the Yerkes Observatory. Green and yellow portions of spectrum. Wave-lengths in millionths of a millimetre

"It is not difficult to select a long list of well-known stars which cannot be far removed from nebular conditions. These are the stars containing both the Huggins and the Pickering series of bright hydrogen lines, the bright lines of helium, and a few others not yet identified. Gamma Argus and Zeta Puppis are of this class. Another is DM +30.3639°, which is actually surrounded with a spherical atmosphere of hydrogen some five seconds of arc in diameter. A little further removed from the nebular state are the stars containing both bright and dark hydrogen lines—caught, so to speak, in the act of changing from bright-line to dark-line stars. Gamma Cassiopeiæ, Pleione, and My Centauri are examples. Closely related to the foregoing are the helium stars. Their absorption lines include the Huggins hydrogen series complete, a score or more of the conspicuous helium lines, frequently a few of the Pickering series, and usually some inconspicuous metallic lines. The white stars in Orion and in the Pleiades are typical of this age.

"The assignment of the foregoing types to an early place in stellar life was first made upon the evidence of the spectroscope. The photographic discovery of nebulous masses in the regions of a large proportion of the bright-line and helium stars affords extremely strong confirmation of their youth. Who that has seen the nebulous background of Orion (Fig. 51) or the remnants of nebulosity in which the individual stars of the Pleiades (Fig. 52) are immersed can doubt that the stars in these groups are of recent formation?

"With the lapse of time, stellar heat radiates into space, and, so far as the individual star is concerned, is lost. On the other hand, the force of gravity on the surface strata increases. The inevitable contraction is accompanied by increasing average temperature. Changes in the spectrum are the necessary consequence. The second hydrogen series vanishes, the ordinary hydrogen absorption is intensified, the helium lines become indistinct, and calcium and iron absorptions begin to assert themselves. Vega and Sirius are conspicuous examples of this period. Increasing age gradually robs the hydrogen lines of their importance, the H and K lines broaden, the metallic lines develop, the bluish-white color fades in the direction of the yellow, and, after passing through types exemplified by many well-known stars, the solar stage is reached. The reversing layer in solar stars represents but four or five hydrogen lines of moderate intensity; the calcium lines are commandingly permanent, and some twenty thousand metallic lines are visible. The solar type seems to be near the summit of stellar life. The average temperature of the mass must be nearly a maximum; for the low density indicates a constitution that is still gaseous [compare Chapter VII.].

"Passing time brings a lowering of the average temperature. The color passes from yellow to red, in consequence of lower radiation, temperature, and increasing general absorption by the atmosphere. The hydrogen lines become indistinct, metallic absorption remains permanent, and broad absorption bands are introduced. In one type (Secchi's Type III.), of which Alpha Herculis is an example, these bands are of unknown origin. In another class (Secchi's Type IV.), illustrated by the star 19 Piscium, they have been definitely identified as of carbon origin.

"There is scarcely room for doubt that these types of stars (Type IV.) are approaching the last stages of stellar development. Surface temperatures have been lowered to the point of permitting more complex chemical combinations than those in the sun.

"Secchi's Type III. includes the several numbered long-period variable stars of the Mira Ceti class, whose spectra at maximum brilliancy show several bright lines of hydrogen and other chemical elements.[1] It is significant that the dull-red stars are all very faint; there are none brighter than magnitude 5.5. Their effective radiatory power is undoubtedly very low."

The state of evolution, which succeeds that character-

---

[1] This circumstance indicates that the red color of these stars, as we have already remarked with regard to Mira Ceti, is not to be traced back to a low temperature, but rather to the dust surrounding them. The most extraordinary brightness of some stars, like Arcturus and Betelgeuse, which are redder than the sun, and whose spectra, according to Hale, resemble those of the sun-spots, presuppose a very high temperature. The characteristic lines of their spectra are produced by the relatively cool vapors of their outer portions.

ORIGIN OF NEBULÆ

ized as the Secchi Type IV., may be elucidated with the aid of the examples of Jupiter and the earth, with which we are more familiar. These planets would be invisible if they were not shining in borrowed light. Jupiter has not advanced so far as the earth. The specific gravity of Jupiter is somewhat lower than that of the sun (1.27 against 1.38), and, apart from the clouds in its atmosphere, this planet is probably altogether in a gaseous condition, while the earth, with its mean density of 5.52, possesses a solid cold crust, enclosing its incandescent interior. This state of the earth corresponds to the last stage in the evolution of the stars.

Of the streams of gaseous matter which are ejected when stars collide with one another, the metallic vapors are rapidly condensed by cooling; only helium and hydrogen will remain in the gaseous condition and form nebular masses about the central body. These nebulæ yield bright lights. Their luminosity is due to the negative particles which are sent to them by the radiation pressure of near stars, and especially by the central bodies of the nebula.

With the new stars which have so far been observed, this pressure of radiation soon diminishes, and the nebular light likewise decreases in such cases. In other instances, as with the stars characterized by bright hydrogen and helium lines, the radiation of the central body or stars in their vicinity seems to be maintained at full force for long periods.

The nebulous accumulations of helium and hydrogen will gradually escape and be condensed in near-by stars under the formation of "explosive" compounds. The tendency to enter into combination seems to be strongest in the case of helium; it disappears first from the stellar atmosphere. That helium enters into compounds at high temperatures seems to follow from the researches of Ramsay, Cooke, and Kohlschütter.

Hydrogen will afterwards be absorbed, and the light of the central body will then show the predominating occurrence of the vapors of calcium and of other metals in its atmosphere. Simultaneously with these, chemical compounds will be noticed, among which the carbon compounds will play an important part—in the outer portions of the sun-spots, in the stars of the Secchi Type IV., as well as in the gaseous envelopes of the comets.[1]

Finally a crust will form. The star is extinct.

[1] The presence of carbon bands in the spectrum need not be taken as a mark of low temperature. Crew and Hale have observed that these bands gradually vanished from an arc spectrum as the temperature was lowered by decreasing the current intensity.

VII

THE NEBULAR AND THE SOLAR STATES

WE will now proceed to a more intimate consideration of the chemical and physical conditions which probably characterize the nebulæ in distinction from the suns. These properties differ in many respects essentially from those which we are accustomed to associate with matter as investigated by us, which may, from this point of view, be styled relatively concentrated.

The differences must be fundamental. For the motto of Clausius, which comprises the sum of our knowledge of the nature of heat, cannot apply to nebulæ. This motto reads:

"The energy of the universe is constant. The entropy of the universe tends to a maximum."

Everybody understands what is meant by energy. We know energy in many forms. The most important are: energy of position (a heavy body has larger energy by virtue of its having been raised to a certain height above the surface of the earth than when it is lying on the surface); energy of motion (a discharged rifle-bullet has an energy which is proportional to the mass of the bullet and to the square of its velocity); energy of heat, which is regarded as the energy of the motion of the smallest particles of a body; electrical energy, such as can, for instance, be stored in an accumulator battery, and which, like all other modifications of energy, may be converted into energy of heat; and chemical energy, such as is displayed by a mixture of eight grammes of oxygen with one gramme of hydrogen, which can be transformed into water under a strong evolution of heat. When we say that the energy of a system to which energy is not imparted from outside is constant, we merely mean that the different forms of energy of the separate parts of this system may be transformed into other forms of energy, but that the sum total of all the energies must always remain unchanged. According to Clausius this law is valid throughout the infinite space of the universe.

By entropy we understand the quantity of heat of a body divided by its absolute temperature. If a quantity of heat, of Q calories, of a body at a temperature of 100° (absolute temperature, 373°) passes over to another body of 0° (absolute temperature, 273°), the total entropy of the two will have been decreased by $\frac{Q}{373}$, and increased by $\frac{Q}{273}$. As the latter quantity is the greater, the entropy of the whole will have increased. By itself, we know, heat always passes, either by radiation or by conduction, from bodies of higher temperature to bodies of lower temperature. That evidently implies an increase in entropy, and it is in agreement with the law of Clausius that entropy tends to increase.

The most simple case of heat equilibrium is that in which we place a number of bodies of unequal temperatures in an enclosure which neither receives heat from outside nor communicates heat to the outside. In some way or other, usually by conduction or radiation, the heat will pass from the warmer to the colder bodies, until at last equilibrium ensues and all the bodies have the same temperature. According to Clausius, the universe tends to that thermal equilibrium. If it be ever

THE NEBULAR AND THE SOLAR STATES

attained, all sources of motion, and hence of light, will have been exhausted. The so-called "heat-death" (Wärmetod) will have come.

If Clausius were right, however, this heat-death, we may object, should already have occurred in the infinitely long space of time that the universe has been in existence. Or we might argue that the world has not yet been in existence sufficiently long, but that, anyhow, it had a beginning. That would contradict the first part of the law of Clausius, that the energy of the universe is constant; for in that case all the energy would have originated in the moment of creation. That is quite inconceivable, and we must hence look for conditions for which the entropy law of Clausius does not hold.

The famous Scotch physicist Clerk-Maxwell has conceived of such a case. Imagine a vessel which is divided by a partition into two halves, both charged with a gas of perfectly uniform temperature. Let the partition be provided with a number of small holes which would not allow more than one gas molecule to pass at a time. In each hole Maxwell places a small, intelligent being (one of his "demons"), which directs all the molecules which enter into the hole, and which have a greater velocity than the mean velocity of all the molecules, to the one side, and which sends to the other side all the molecules of a smaller velocity than the average.[1] All the undesirable molecules the demon bars by means of a little flap. In this way all the molecules of a velocity greater than the average may be collected in the one compartment, and all the molecules of a lesser velocity in the other com-

---

[1] The kinetic theory of gases imagines all the molecules of a gas to be in constant motion. The internal pressure of the gas depends upon the mean velocity of the particles; but some particles will move at a greater, and some at a smaller velocity than the average.—H.B.

partment. In other words, heat—for heat consists in the movements of molecules—will pass from the one constantly cooling side to the other, which is constantly raising its temperature, and which must therefore become warmer than the former.

In this instance heat would therefore pass from a colder to a warmer body, and the entropy would diminish.

Nature, of course, does not know any such intelligent beings. Nevertheless, similar conditions may occur in celestial bodies in the gaseous state. When the molecules of gas in the atmosphere of a celestial body have a sufficient velocity—which in the case of the earth would be 11 km. (7 miles) per second—and when they travel outward into the most extreme strata, they may pass from the range of attraction out into infinite space, after the manner of a comet, which, if endowed with sufficient velocity when near the sun, must escape from the solar system. According to Stoney, it is in this way that the moon has lost its original atmosphere. This loss of gas is certainly imperceptible in the case of our sun and of large planets like the earth. But it may play an important part in the household of the nebulæ, where all the radiation from the hot celestial bodies is stored up, and where, owing to the enormous distances, the restraining force of gravity is exceedingly feeble. Thus the nebulæ will lose their most rapid molecules from their outer portions, and they will therefore be cooling in these outer strata. This loss of heat is compensated by the radiation from the stars. If, now, there were only nebulæ of one kind in the whole universe, those escaped molecules would finally land on some other nebula, heat equilibrium would thus be established between the different nebulæ, and the "heat-death" be realized. But we have already remarked that the nebulæ enclose many immigrated celestial bodies, which are able to condense the gases from their neighborhood, and which thereby assume a higher temperature. The lost molecules of gases may also stray into the vast atmosphere of these growing stars, and the condensation will then be hastened under a continuous lowering of the entropy. By such processes the clock-work of the universe may be maintained in motion without running down.

About the bodies which have drifted into nebulæ, and about the remnants of new stars which lie inside the nebulæ, the gases will thus collect which had formerly been scattered through the outer portions of the nebula. These gases originate from the explosive compounds which had been stored in the interior of the new stars. Hydrogen and helium are, most likely, the most important of these; for they are the most difficult to be condensed, and can exist in notable quantities at extremely low temperatures, such as must prevail in the outermost portions of the nebulæ, in which gases of other substances would be liquefied. Even if the nebulæ had an absolute temperature of 50° ($-223°$ C.), the vapor of the most volatile of all the metals, mercury, would even in the saturated state be present in such a small quantity that a single gramme would occupy the space of a cube whose side would correspond to about two thousand light-years —that is to say, to 450 times the distance of the earth from the nearest fixed star. One gramme of sodium, likewise a very volatile metal, and of a comparatively high importance in the constitution of the fixed stars, would fill the side of a cube that would be a thousand million times as large. Still more inconceivable numbers result for magnesium and iron, which are very frequent constituents of fixed stars, and which are less volatile than the just-mentioned metals. We thus recognize the strongly selective action of the low temperatures upon all the substances which are less difficult to condense than helium and hydrogen. As we now know that there is another substance in the nebulæ, which has been designated nebulium, and which is characterized by two spectral lines not found in any terrestrial substance, we must conclude that this otherwise unknown element nebulium must be almost as difficult to condense as hydrogen and helium. Its boiling-point will probably lie below 50° absolute, like that of those gases.

That hydrogen and helium, together with nebulium, alone seem to occur in the vastly extended nebulæ is probably to be ascribed to their low boiling-points. We need not look for any other explanation. The supposition of Lockyer that all the other elements would be transformed into hydrogen and helium at extreme rarefaction is quite unsupported.

In somewhat lower strata of the nebula, where its shape resembles a disk, other not easily condensable substances, such as nitrogen, hydro-carbons of simple composition, carbon monoxide, further, at deeper levels, cyanogen and carbon dioxide, and, near the centre, sodium, magnesium, and even iron may occur in the gaseous state. These less volatile constituents may exist as dust in the outermost strata. This dust would not be revealed to us by the spectroscope. In the strongly developed spiral nebulæ, however, the extreme layers, which seem to hide the central body, appear to be so attenuated that the dust floating in them is not able to obscure the spectrum of the metallic gases. The spectrum of the nebula then resembles a star spectrum, because the deepest strata contain incandescent layers of dust clouds, whose light is sifted by the surrounding masses of gases.

It has been observed that the lines of the different elements are not uniformly distributed in the nebulæ. Thus Campbell observed, for instance, when investigating a small planetary nebula in the neighborhood of the great Orion nebula, that the nebulium had not the same distribution as the hydrogen. The nebulium, which was concentrated in the centre of the nebula, probably has a higher boiling-point than hydrogen, therefore, and occurs in noticeable quantities in the inner, hotter parts of the nebula. Systematic investigations of this kind may help us to a more perfect knowledge of the temperature relations in these peculiar celestial objects.

Ritter and Lane have made some interesting calculations on the equilibrium in a gaseous celestial body of so low a density that the law of gases may be applied to it. That is only permissive for gases or for mixtures of gases whose density does not exceed one-tenth of that of water or one-fourteenth of the actual density of the sun. The pressure in the central portions of such a mass of gas would, of course, be greater than the pressure in the outer portions, just as the pressure rises as we penetrate from above downward into our terrestrial atmosphere. If we imagine a mass of the air of our atmosphere transferred one thousand metres higher up, its volume will increase and its temperature will fall by 9.8° C. (18° F.). If there were extremely violent vertical convection currents in the air, its temperature would diminish in this manner with increasing altitude; but internal radiation tends to equalize these temperature differences. The following calculation by Schuster concerning the conditions of a mass of gas of the size of the sun is based on Ritter's investigation. It has been made under the hypothesis that the thermal properties of this mass of gas are influenced only by the movements in it, and not by radiation. The calculation is applied to a star which has the same mass as the sun (1.9×10^{33} grammes, or 324,000 times the mass of the earth), and a radius of about ten times that of the sun ($10 \times 690,000$ km.), whose mean density would thus be 1000 times smaller than that of the sun, or 0.0014 times the density of water at 4° C. In the following table the first column gives the distance of a point from the centre of the star as a fraction of its radius; the density (second column) is expressed in the usual scale, water being the unit; pressures are stated in thousands of atmospheres, temperatures in thousands of degrees Centigrade. The temperature will vary proportionately to the molecular weight of the gas of which the star consists; the temperatures, in the fourth column of the table, concern a gas of molecular weight 1—that is to say, hydrogen gas dissociated into atoms, as it will be undoubtedly on the sun and on the star. If the star should consist of iron, we should have to multiply these latter numbers by 56, the molecular weight of iron; the corresponding figures will be found in the fifth column.

Distance from centre	Density	Pressure in 10^3 atmospheres	Temperature in 10^{30} Cent.	
			Hydrogen	Iron
0	0.00844	852	2460	137,500
0.1	0.00817	807	2406	134,600
0.2	0.00739	683	2251	126,100
0.3	0.00623	513	2007	112,400
0.4	0.00488	342	1707	95,600
0.5	0.00354	200	1377	77,100
0.6	0.00233	100	1043	58,400
0.7	0.00136	40	728	48,800
0.8	0.00065	12	445	24,900
0.9	0.00020	1.7	202	11,300
1.0	0.00000	0	0	0

Schuster's calculation was really made for the sun — that is to say, for a celestial body whose diameter is ten times smaller, and whose specific gravity is therefore a thousand times greater than the above-assumed values. According to the laws of gravitation and of gases, the pressure must there be 10,000 times greater, and the temperature ten times higher, than those in our table. The density of the interior portions would, however, become far too large to admit of the application of the gas laws. I have therefore modified the calculations so as to render them applicable to a celestial body of ten times the radius of the sun or of 1080 times the radius of the earth; the radius would then represent one-twenty-second of the distance from the centre of the sun to the earth's orbit, and the respective celestial body would have very small dimensions indeed if compared to a nebula.

The extraordinarily high pressure in the interior portions of the celestial body is striking; this is due to the great mass and to the small distances. In the centre of the sun the pressure would amount to 8520 million atmospheres, since the pressure increases inversely as the fourth power of the radius. The pressure near the centre of the sun is, indeed, almost of that order. If the sun were to expand to a spherical planetary nebula of a thousand times its actual linear dimensions (when it would almost fill the orbit of Jupiter), the specific gravity at its centre would be diminished to one-millionth of the above-mentioned value—that is to say, matter in this nebula would not, even at the point of greatest concentration, be any denser than in the highly rarefied vacuum tubes which we can prepare at ordinary temperatures. The pressure would likewise be greatly diminished—namely, to about six millimetres only, near the centre of the gaseous mass. The temperature, however, would be rather high near the centre—namely, 24,600° C., if the nebula should consist of atomatic hydrogen, and fifty-six times as high again if consisting of iron gas. Such a nebula would restrain gases with 1.63 times the force which the earth exerts. Molecules of gases moving outward with a velocity of about 18 km. (11 miles) per second would forever depart from this atmosphere.

The estimation of the temperature in such masses of gases is certainly somewhat unreliable. We have to presume that neither radiation nor conduction exert any considerable influence. That might be permitted for conduction; but we are hardly justified in neglecting radiation. The temperatures within the interior of the nebula will, therefore, be lower than our calculated values. It is, however, difficult to make any definite allowance for this factor.

If the mass of the celestial body should not be as presumed—for instance, twice as large—we should only have to alter the pressure and the density of each layer in the same proportion, and thus to double the above values. The temperature would remain unchanged. We are hence in a position to picture to ourselves the state of a nebula of whatever dimensions and mass.

Lane has proved, what the above calculations also indicate, that the temperature of such nebula will rise when it contracts in consequence of its losing heat. If heat were introduced from outside, the nebula would expand under cooling. A nebula of this kind presumably loses heat and gradually raises its own temperature until it has changed into a star, which will at first have an atmosphere of helium and of hydrogen like that of the youngest stars (with white light). By-and-by, under a further rise of temperature, the extremely energetic chemical compounds will be formed which characterize the interior of the sun, because helium and hydrogen—which were liberated when the nebula was re-formed and which

THE NEBULAR AND THE SOLAR STATES

dashed out into space—will diffuse back into the interior of the star, where they will be bound under the formation of the compounds mentioned. The atmosphere of hydrogen and of helium will disappear (helium first), the star will contract more and more, and the pressure and the convection currents in the gases will become enormous. There will be a strong formation of clouds in the atmosphere of the star, which will gradually become endowed with the properties which characterize our sun. The sun behaves very differently from the gaseous nebulæ for which the calculations of Lane, Ritter, and Schuster hold. For when the contraction of a gas shall have proceeded to a certain limit, the pressure will increase in the ratio 1 : 16, while the volume will decrease in the ratio 8 : 1, provided there be no change in the temperature. When the gas has reached this point and is still further compressed, the temperature will remain in steady equilibrium. At still higher pressures, however, the temperature must fall if equilibrium is to be maintained. According to Amagat, this will occur at 17° C. (290° absolute) in gases like hydrogen and nitrogen, which at this temperature are far above their critical points, and at a pressure of 300 or 250 atmospheres. When the temperature is twice as high on the absolute scale, or at 307° C., twice the pressure will be required.

We can now calculate when our nebula will pass through this critical stage, to which a lowering of the temperature must succeed. Accepting the above figures, we find that half the mass of the nebula will fill a sphere of a radius 0.53 of that of the nebula. If the mass were everywhere of the same density, half of it would fill a sphere of 0.84 of this radius. When will the interior mass cross the boundary of the above stage, while the exterior portions still remain below this stage? That will be at about the time when the nebula in its totality will pass through its maximum temperature. We will now base our calculations on the temperatures which apply to iron in the gaseous state; for in the interior of the nebula the mean molecular weight will at least be 56 (that of iron). We shall find that the pressure at the distance 0.53 will be about 177,000 atmospheres, and the temperature approximately 71 million degrees—*i.e.*, 245,000 times higher than the absolute temperature in the experiments of Amagat. The specified stage will then be reached when the pressure will be 245,000 times as large as 250 atmospheres—viz., 61 million atmospheres. As, now, the pressure is only 177,000 atmospheres, our nebula will yet be far removed from that stage at which cooling will set in. We can easily calculate that this will take place when the nebula has contracted to a volume about three times that of our sun. The assertion which is so often made that the sun might possibly attain higher temperatures in the future is unwarranted. This celestial body has long since passed through the culminating-point of its thermal evolution, and is now cooling. As the temperatures which Schuster deduced were no doubt much too high, the cooling must, indeed, have set in already in an earlier stage. But stars like Sirius, whose density is probably not more than one per cent. of the solar density, are probably still in a rising-temperature stage. Their condition approximates that of the mass of gas of our example.

The planetary nebulæ are vastly more voluminous. The immense space which these celestial bodies may occupy will be understood from the fact that the largest among them, No. 5 in Herschel's catalogue, situated near the star B in the Great Bear, has a diameter of 2.67 seconds of arc. If it were as near to us as our nearest

THE NEBULAR AND THE SOLAR STATES

star neighbor, its diameter would yet be more than three times that of the orbit of Neptune; doubtless it is many hundreds of times larger. This consideration furnishes us with an idea of the infinite attenuation in such structures. In their very densest portions the density cannot be more than one-billionth of the density of the air. In the outer portions of such nebulæ the temperature must also be exceedingly low; else the particles of the nebula could not be kept together, and only hydrogen and helium can occur in them in the gaseous state.

Yet we may regard the density and temperature of such celestial bodies as gigantic by comparison with those of the gases in the spirals of the nebulæ. There never is equilibrium in these spirals, and it is only because the forces in action are so extraordinarily small that these structures can retain their shapes for long periods without noticeable changes. It is, probably, chiefly in those parts in which the cosmical dust is stopped in its motion that meteorites and comets are produced. The dust particles wander into the more central portions of the nebulæ, into which they penetrate deeply, owing to their relatively large mass, to form the nuclei for the growth of planets and moons. By their collisions with the masses of gases which they encounter, they gradually assume a circular movement about the axis of rotation of the nebula. In this rotation they condense portions of the gases on their surface, and hence acquire a high temperature—which they soon lose again, however, owing to the comparatively rapid radiation.

So far as we know, spiral nebulæ are characterized by continuous spectra. The splendor of the stars within them completely outshines the feeble luminosity of the nebula. The stars in them are condensation products and undoubtedly in an early stage of their existence;

they may therefore be likened to the white stars, like the new star in Perseus and the central star in the ring nebula of the Lyre. Nevertheless, it has been ascertained that the spectrum of the Andromeda nebula has about the same length as that of the yellow stars. That may be due to the fact that the light of the stars in this nebula, which we only seem to see from the side, is partly extinguished by dust particles in its outer portion, as was the case with the light of the new star in Perseus during the period of its variability.

Our considerations lead to the conclusion that there is rotating about the central body of the nebula an immense mass of gas, and that outside this mass there are other centres of condensation moving about the central body together with the masses of gas concentrated about them. Owing to the friction between the immigrated masses and the original mass of gas which circulated in the equatorial plane of the central body, all these masses will keep near the equatorial plane, which will therefore deviate little from the ecliptic. We thus obtain a proper planetary system, in which the planets are surrounded by colossal spheres of gas like the stars in the Pleiades (Fig. 52). If, now, the planets have very small mass by comparison with the central body—as in our solar system—they will be cooled at an infinitely faster rate than the sun. The gaseous masses will soon shrink, and the periods of rotation will be shortened; but for those planets, at least, which are situated near the centre, these periods will originally differ little from the rotation of the central body. The dimensions of the central body will always be very large, and the planets circulating about it will produce very strong tidal effects in its mass. Its period of rotation will be shortened, while the orbital rotation of the planets will tend to be-

THE NEBULAR AND THE SOLAR STATES

come lengthened. Thus the equilibrium is disturbed; it is re-established again, because the planet is, so to say, lifted away from the sun, as G. H. Darwin has so ingeniously shown with regard to the moon and the earth. Similar relations will prevail in the neighborhood of those planets which will thus become provided with moons. Hence we understand the peculiar fact that all the planets move almost in the same plane, the so-called ecliptic, and in approximately circular orbits; that they all move in the same direction, and that they have the same direction of rotation in common with their moons and with the central body, the sun. It is only the outermost planets, like Uranus and Neptune, in whose cases the tidal effects were not of much consequence, that form exceptions to this rule.

In explanation of these phenomena various philosophers and astronomers have advanced a theory which is known as the Kant-Laplace theory, after its most eminent advocates. Suggestions pointing in the same direction we find in Swedenborg (1734). Swedenborg assumed that our planetary system had been evolved under the formation of vortices from a kind of "chaos solare," which had acquired a more and more energetic circulating motion about the sun under the influence of internal forces, possibly akin to magnetic forces. Finally a ring had been thrown off from the equator, and had separated into fragments, out of which the planets had been formed.

Buffon introduced gravitation as the conservational principle. In an ingenious essay, "Formation des Planètes" (1745), he suggests that the planets may have been formed from a "stream" of matter which was ejected by the sun when a comet rushed into it.

Kant started from an original chaos of stationary dust, which under the influence of gravitation arranged itself as a central body, with rings of dust turning around it; the rings, later on, formed themselves into planets. The laws of mechanics teach, however, that no rotation can be set up in a central body, which is originally stationary, by the influence of a central force like gravitation. Laplace, therefore, assumed with Swedenborg that the primeval nebula from which our solar system was evolved had been rotating about the central axis. According to Laplace, rings like those of Saturn would split off, as such a system contracted, and planets and their moons and rings would afterwards be formed out of those rings. It is generally believed at present, however, that only meteorites and small planets, but not the larger planets, could have originated in this way. We have, indeed, such rings of dust rotating about Saturn, the innermost more rapidly, the outer rings more slowly, just as they would if they were crowds of little moons.

Many further objections have later been raised against the hypothesis of Laplace, first by Babinet, later especially by Moulton and Chamberlin. In its original shape this hypothesis would certainly not appear to be tenable. I have therefore replaced it by the evolution thesis outlined above. It is rather striking that the moons of the outermost planets, Neptune and Uranus, do not move in the plane of the ecliptic, and that their moons further describe a "retrograde" movement—that is to say, they move in the direction opposite to that conforming to the theory of Laplace. The same seems to hold for the moon of Saturn, which was discovered in 1898 by Pickering. All these facts were, of course, unknown to Laplace in 1776; and if he had known them he would scarcely have advanced his thesis in the garb in which he offered it. The explanation of these facts does not cause any difficulty. We may assume that the matter in the outer portions of the primeval nebula was so strongly attenuated that the immigrating planet did not attain a sufficient volume to have the large common rotation in the equatorial plane of the sun impressed upon it by the tidal effects. Charged only with the small mass of matter which they met on their road, the planet and its moon, on the contrary, remained victorious in the limited districts in which they were rotating. Only the slow orbital movement about the central body was influenced, and that adapted itself to the common direction and the circular orbit. It is not inconceivable that there may be, farther out in space, planets of our solar system, unknown to us, moving in irregular paths like the comets. The comets, Laplace assumed, probably immigrated at a later period into our solar system when the condensation had already advanced so far that the chief mass of the nebular matter had disappeared from interplanetary space.

Chamberlin and Moulton have attempted to show that the difficulties of the hypothesis of Laplace may be obviated by the assumption that the solar system has evolved from a spiral nebula, into which strange bodies intruded which condensed the nebular mass of their surroundings upon themselves. We have pointed out examples of how the nebula seems to vanish in the vicinity of the stars, which would correspond to growing planets, located in nebulæ.

In concluding this consideration, we may draw a comparison between the views which were still entertained a short time ago and the views and prospects which the discoveries of modern days open to our eyes.

Up to the beginning of this century the gravitation of Newton seemed to rule supreme over the motions and over the development of the material universe. By virtue of this gravitation the celestial bodies should tend to draw together, to unite in ever-growing masses. In the infinite space of past time the evolution should have proceeded so far that some large suns, bright or extinct, could alone persist. All life would be impossible under such conditions.

And yet we discern in the neighborhood of the sun quite a number of dark bodies, our planets, and we may surmise that similar dark companions or satellites exist in the vicinity of other suns and stars; for we could not understand the peculiar to-and-fro motions of those stars on any other view. We further observe that quite a number of small celestial bodies rush through space in the shapes of meteorites or shooting-stars which must have come to us from the most remote portions of the universe.

The explanation of these apparent deviations from what we may regard as a necessary consequence of the exclusive action of gravity will be found under two heads—in the action of the mechanical radiation pressure of light, and in the collisions between celestial bodies. The latter produce enormous vortices of gases about nebular structures in the gaseous condition; the radiation pressure carries cosmical dust into the vortices, and the dust collects into meteorites and comets and forms, together with the condensation products of the gaseous envelope, the planets and the moons accompanying them.

The scattering influence of the radiation pressure therefore balances the tendency of gravitation to concentrate matter. The vortices of gases in the nebulæ only serve to fix the position of the dust, which is ejected from the suns through the action of the radiation pressure.

The masses of gas within the nebulæ form the most important centres of concentration of the dust which is eject-

THE NEBULAR AND THE SOLAR STATES

ed from the sun and stars. If the world were limited, as people used to fancy—that is to say, if the stars were crowded together in a huge heap, and only infinite, empty space outside of this heap, the dust particles ejected from the suns during past ages by the action of the radiating pressure would have been lost in infinite space, just as we imagined that the radiated energy of the sun was lost.

If that were so, the development of the universe would long since have come to an end, to an annihilation of all matter and of all energy. Herbert Spencer, among others, has explained how thoroughly unsatisfactory this view is. There must be cycles in the evolution of the universe, he has emphasized. That is manifestly indispensable if the system is to last. In the more rarefied, gaseous, cold portions of the nebulæ we find that part of the machinery of the universe which checks the waste of matter and, still more, the waste of force from the suns. The immigrating dust particles have absorbed the radiation of the sun and impart their heat to the separate particles of the gases with which they collide. The total mass of gas expands, owing to this absorption of heat, and cools in consequence. The most energetic molecules travel away, and are replaced by new particles coming from the inner portions of the nebulæ, which are in their turn cooled by expansion. Thus every ray emitted by a sun is absorbed, and its energy is transferred, through the gaseous particles of the nebulæ, to suns that are being formed and which are in the neighborhood of the nebula or in its interior portions. The heat is hence concentrated about centres of attraction that have drifted into the nebula or about the remnants of the celestial bodies which once collided there. Thanks to the low temperature of the nebula, the matter can again accumulate, while the radiation pressure, as Poynting has shown, will suffice to keep bodies apart if their temperature is 15° C., their diameter 3.4 cm., and their specific gravity as large as that of the earth, 5.5. At the distance of the orbit of Neptune, where the temperature is about 50° absolute and approximates, therefore, that of a nebula, this limit of size is reduced to nearly one millimetre. It has already been suggested (compare page 153) that capillary forces, which would prevail under the co-operation of the gases condensed upon the dust grains, rather than gravity, play a chief part in the accumulation and coalescence of the small particles. In the same manner as matter is concentrated about centres of attraction energy may be accumulated there in contradiction to the law of the constant increase of entropy.

During this conservational activity the layers of gas are rapidly rarefied, to be replaced by new masses from the inner parts of the nebula, until this centre is depleted, and the nebula has been converted into a star cluster or a planetary system which circulates about one or several suns. When the suns collide once more new nebulæ are created.

The explosive substances, consisting probably of hydrogen and helium (and possibly of nebulium), in combination with carbon and metals, play a chief part in the evolution from the nebular to the stellar state, and in the formation of new nebulæ after collisions between two dark or bright celestial bodies. The chief laws of thermodynamics lead to the assumption that these explosive substances are formed during the evolution of the suns and are destroyed during their collisions. The enormous stores of energy concentrated in these bodies perform, in a certain sense, the duty of powerfully acting fly-wheels interposed in the machinery of the

THE NEBULAR AND THE SOLAR STATES

universe in order to regulate its movements and to make certain that the cyclic transition from the nebular to the star stage, and vice versa, will occur in a regular rhythm during the immeasurable epochs which we must concede for the evolution of the universe.

By virtue of this compensating co-operation of gravity and of the radiation pressure of light, as well as of temperature equalization and heat concentration, the evolution of the world can continue in an eternal cycle, in which there is neither beginning nor end, and in which life may exist and continue forever and undiminished.

VIII

THE SPREADING OF LIFE THROUGH THE UNIVERSE

We have just recognized the probability of the assumption that solar systems have been evolved from nebulæ, and that nebulæ are produced by the collision of suns. We likewise consider it probable that there circulate about the newly formed suns smaller celestial bodies which cool more rapidly than the central sun. When these satellites have provided themselves with a solid crust, which will partly be covered by water, they may, under favorable conditions, harbor organic life, as the earth and probably also Venus and Mars do. The satellites would thereby gain a greater interest for us than if we had to imagine them as consisting entirely of lifeless matter.

The question naturally arises whether we may believe that life can really originate on a celestial body as soon as circumstances are favorable for its evolution and propagation. This question will occupy us in this last chapter.

Men have been pondering over these problems since the remotest ages. All living beings, past ages recognized, must have been generated and they had to die after a certain shorter or longer life. Somewhat later, and yet still in a very early epoch, experience must have taught men that organisms of one kind can only generate other organisms of the same kind; that the species are invariable, as we now express it. The idea was that all species originally came from the hands of the Creator endowed with their present qualities. This view may still be said to represent the general or "orthodox" doctrine. This view has also been called the Linnæan thesis, because Linné, in the fifth edition of his *Genera Plantarum*, adheres to it strictly: "Species tot sunt, quot diversas formas ab initio produxit Infinitum Ens, quae deinde formae secundum generationis inditas leges produxere plures, at sibi semper similes, ut species nunc nobis non sint plures quam fuerunt ab initio." Which we may render: "There are as many different kind of species as the Infinite Being has created different forms in the beginning. These forms have later engendered other beings according to the laws of inheritance, always resembling them, so that we have at the present time not any more species than there were from the beginning." Time was ripe, however, even then for a less rigid conception of nature, more in accordance with our present views. The first foundations of the theory of evolution in the biological sciences were laid by Lamarck (in 1794), Treviranus (in 1809), Goethe and Oken (in 1820). But a reaction set in. Cuvier and his authority forced public opinion back to the ancient stand-point. In his view the now extinct species of past geological epochs had been destroyed by natural revolutions, and new species had again been generated by a new act of the Creator.

Within the last few decades, however, the general belief has rapidly been revolutionized, and the theory of evolution, especially since the immortal Charles Darwin came forth with his epoch-making researches, now meets with universal acceptance.

According to this theory the species adapt themselves in the course of time to their surroundings, and the changes may become so great that a new species may be considered to have originated from an old species. The researches of De Vries have, within quite recent times, further accentuated this view, so that we now concede cases to be extant where new species spring forth from old ones under our very eyes. This thesis has become known as the theory of mutation.

At the present time we accordingly imagine that living organisms, such as we see around us, have all descended from older organisms, rather unlike them, of which we still find traces and remnants in the geological strata which have been deposited during past ages. From this stand-point all living organisms might possibly have originated from one single, most simple organism. How that was generated still remains to be explained.

The common view, to which the ancients inclined, is that the lower organisms need not necessarily have originated from seeds. It was noticed that some low-type organisms, larvæ, etc., took rise in putrid meat; Vergil describes this in his *Georgicas*. It was not until the seventeenth century that this belief was disproved by many experiments, among others by those of Swammerdam and Leuwenhoek. The thesis of the so-called "Generatio spontanea" once more blossomed into new life upon the discovery of the so-called infusoria, the small animal organisms which seem to arise spontaneously in infusions and concoctions. Spallanzani, however, demonstrated in 1777 that when the infusions, and the vessel containing them, as well as the air above them, were heated to a sufficiently high temperature to kill all the germs present, the infusions would remain sterile, and no living organisms could develop in them. To this fact we owe our ordinary methods of making preserves. It is true that objections were raised against this demonstration. The air, it was objected, is so changed by heating that subsequent development of minute organisms is rendered impossible. But this last objection was refuted by the chemists Chevreul and Pasteur, as well as by the physicist Tyndall in the sixties and seventies of the past century. These scientists demonstrated that no organisms are produced in air which is freed from the smallest germs by some other means than heating—*i.e.*, by filtration through cotton-wool. The researches of Pasteur, in particular, and the methods of sterilization which are based upon them and which are applied every day in bacteriological laboratories, have more and more forced the conviction upon us that a germ is indispensable for the origination of life.

And yet eminent scientists take up the pen again and again in order to demonstrate the possibility of the "Generatio spontanea." In this they do not rely upon the safe methods of natural science, but they proceed on philosophical lines of argument. Life, they suggest, must once have had a beginning, and we are hence forced to believe that spontaneous generation, even if not realizable under actual conditions, must have once occurred. Considerable interest was excited when the great English physiologist Huxley believed he had discovered in the mud brought up from the very bottom of the sea an albuminoid substance which he called "Bathybius Haeckelii," in honor of the zealous German Darwinist Haeckel. In this bathybius (deep-sea organism) one fancied for a time that the primordial ooze, which had originated from inorganic matter and from which all organisms might have been evolved, and of which Oken had been dreaming, had been discovered. But the more exact researches of the chemist Buchanan demonstrated that the albuminoid substance in this primordial ooze consisted of flocks of gypsum precipitated by alcohol.

People then had recourse to the most fantastic speculations. Life, it was argued, might possibly have had its origin in the incandescent mass of the interior of the earth. At high temperatures organic compounds of cyanogen and its derivatives might be formed which would be the carriers of life (Pflüger). There is, however, little need of our entering into any of these speculations until they have been provided with an experimental basis.

Almost every year the statement is repeated in biological literature that we have at last succeeded in producing life from dead matter. Among the most recent assertions of this kind, the discovery claimed by Butler-Burke has provoked much comment. He asserted that he had succeeded, with the aid of the marvellous substance radium, in instilling life into lifeless matter—namely, a solution of gelatine. Criticism has, however, relegated this statement, like all similar ones, to the realm of fairy tales.

We fully share the opinion which the great natural philosopher Lord Kelvin has expressed in the following words: "A very ancient speculation, still clung to by many naturalists (so much so that I have a choice of modern terms to quote in expressing it), supposes that, under meterological conditions very different from the present, dead matter may have run together or crystallized or fermented into 'germs of life,' or 'organic cells,' or 'protoplasm.' But science brings a vast mass of inductive evidence against this hypothesis of spontaneous generation. Dead matter cannot become living without coming under the influence of matter previously alive.

LIFE THROUGH THE UNIVERSE

This seems to me as sure a teaching of science as the law of gravitation."

Although the latter verdict may be a little dogmatic, it yet demonstrates how strongly many scientists feel the necessity of finding another way of solving the problem. The so-called theory of panspermia really shows a way. According to this theory life-giving seeds are drifting about in space. They encounter the planets, and fill their surfaces with life as soon as the necessary conditions for the existence of organic beings are established.

This view was probably foreshadowed long ago. Definite suggestions in this direction we find in the writings of the Frenchman Sales-Guyon de Montlivault (1821), who assumed that seeds from the moon had awakened the first life on the surface of the earth. The German physician H. E. Richter attempted to supplement the doctrine of Darwin by combining the conception of panspermia with it. Flammarion's book on the plurality of inhabited worlds suggested to Richter the idea that seeds had come from some other inhabited world to our earth. He emphasizes the fact that carbon has been found in meteorites which move in orbits similar to those of the comets which wander about in space; and in this carbon he sees the rests of organic life. There is no proof at all for this latter opinion. The carbon found in meteorites has never exhibited any trace of organic structure, and we may well imagine the carbon—$e.g.$, that which appears to occur in the sun—to be of inorganic origin. Still more fantastic is his idea that organisms floating high in our atmosphere are caught by the attraction of meteorites flying past our planet, and are in this way carried out into universal space and deposited upon other celestial bodies. As the surface of meteorites becomes incandescent in their flight through the atmosphere, any germs which they might possibly have caught would be destroyed; and if, in spite of that, a meteorite should become the conveyor of live germs, those germs would be burned in the atmosphere of the planet on which they descended.

In one point, however, we must agree with Richter. There is logic in his statement that "The infinite space is filled with, or (more correctly) contains, growing, mature, and dying celestial bodies. By mature worlds we understand those which are capable of sustaining organic life. We regard the existence of organic life in the universe as eternal. Life has always been there; it has always propagated itself in the shape of living organisms, from cells and from individuals composed of cells." Man used to speculate on the origin of matter, but gave that up when experience taught him that matter is indestructible and can only be transformed. For similar reasons we never inquire into the origin of the energy of motion. And we may become accustomed to the idea that life is eternal, and hence that it is useless to inquire into its origin.

The ideas of Richter were taken up again in a popular lecture delivered in 1872 by the famous botanist Ferdinand Cohn. The best-known expression of opinion on the subject, however, is that of Sir William Thomson (later Lord Kelvin) in his presidential address to the British Association at Edinburgh in 1871:

"When two great masses come into collision in space, it is certain that a large part of each is melted; but it seems also quite certain that in many cases a large quantity of débris must be shot forth in all directions, much of which may have experienced no greater violence than individual pieces of rock experience in a landslip or in blasting by gunpowder. Should the time when this

LIFE THROUGH THE UNIVERSE

earth comes into collision with another body, comparable in dimensions to itself, be when it is still clothed as at present with vegetation, many great and small fragments carrying seed and living plants and animals would undoubtedly be scattered through space. Hence, and because we all confidently believe that there are at present, and have been from time immemorial, many worlds of life besides our own, we must regard it as probable in the highest degree that there are countless seed-bearing meteoric stones moving about through space. If at the present instant no life existed upon this earth, one such stone falling upon it might, by what we blindly call *natural* causes, lead to its becoming covered with vegetation. I am fully conscious of the many objections which may be urged against this hypothesis. I will not tax your patience further by discussing any of them on the present occasion. All I maintain is that I believe them to be all answerable."

Unfortunately we cannot share Lord Kelvin's optimism regarding this point. It is, in the first instance, questionable whether living beings would be able to survive the violent impact of the collision of two worlds. We know, further, that the meteorite in its fall towards the earth becomes incandescent all over its surface, and any seeds on it would therefore be deprived of their germinating power. Meteorites, moreover, show quite a different composition from that of the fragments from the surface of the earth or a similar planet. Plants develop almost exclusively in loose soil, and a lump of earth falling through our atmosphere would, no doubt, be disintegrated into a shower of small particles by the resistance of the atmosphere. Each of these particles would by itself flash up like a shooting-star, and could not reach the earth in any other shape than that of burned dust.

Another difficulty is that such collisions, which, as we presume, are responsible for the flashing-up of so-called new stars, are rather rare phenomena, so that little likelihood remains of small seeds being transported to our earth in this manner.

The question has, however, entered into a far more favorable stage since the effects of radiation have become understood.

Bodies which, according to the deductions of Schwarzschild, would undergo the strongest influence of solar radiation must have a diameter of 0.00016 mm., supposing them to be spherical. The first question is, therefore: are there any living seeds of such extraordinary minuteness? The reply of the botanist is that the so-called permanent spores of many bacteria have a size of 0.0003 or 0.0002 mm., and there are, no doubt, much smaller germs which our microscopes fail to disclose. Thus, yellow-fever in man, rabies in dogs, the foot-and-mouth disease in cattle, and the so-called mosaic disease—common to the tobacco plant in Netherlandish India, and also observed in other countries—are, no doubt, parasitical diseases; but the respective parasites have not yet been discovered, presumably because they are too minute to be visible under the microscope.[1]

It is, therefore, very probable that there are organisms so small that the radiation pressure of a sun would push them out into space, where they might give rise to life on planets, provided they met with favorable conditions for their development.

We will, in the first instance, make a rough calculation

---

[1] Meanwhile a large number of organisms which are invisible under the ordinary microscope have been rendered visible by the aid of the ultra-microscope, among others the presumable microbe of the foot-and-mouth disease.

of what would happen if such an organism were detached from the earth and pushed out into space by the radiation pressure of our sun. The organism would, first of all, have to cross the orbit of Mars; then the orbits of the smaller and of the outer planets; and, having passed the last station of our solar system, the orbit of Neptune, it would drift farther into infinite space towards other solar systems. It is not so difficult to estimate the time which the smallest particles would require for this journey. Let their specific gravity be that of water, which will very fairly correspond to the facts. The organisms would cross the orbit of Mars after twenty days, the Jupiter orbit after eighty days, and the orbit of Neptune after fourteen months. Our nearest solar system, Alpha Centauri, would be reached in nine thousand years. These calculations have been made under the supposition that the radiation pressure is four times as strong as gravitation, which would be nearly correct according to the figures of Schwarzschild.[1]

These time intervals required for the organisms to reach the different planets of our solar system are not too long for the germs in question to preserve their germinating power. The estimate is more unfavorable in the case of their transference from one planetary system to another, which will require thousands of years. But we shall see further on that the very low temperature of those parts of space (about $-220°$ C.) would suspend the extinction of the germinating power, as it arrests all chemical reactions.

As regards the period during which the germinating

---

[1] The radiation pressure has here been assumed to be somewhat greater than on page 103, because the spores are here regarded as opaque, while the drops of hydrocarbons have been regarded as partially translucid to luminous rays.

power can be preserved at ordinary temperature, we have been told that the so-called "mummy wheat" which had been found in ancient Egyptian tombs was still capable of germination. Critics, however, have established that the respective statements of the Arabs concerning the sources of that wheat are very doubtful. The French scientist Baudoin asserts that bacteria capable of germination were found in a Roman tomb which had certainly remained untouched for eighteen hundred years; but this statement is to be received with caution. It is certain, however, that both seeds of some higher plants and spores of certain bacteria—*e. g.*, anthrax—do maintain their germinating power for several years (say, twenty), and thus for periods which are much longer than those we have estimated as necessary for their transference to our planet.

On the road from the earth the germs would for about a month be exposed to the powerful light of the sun, and it has been demonstrated that the most highly refrangible rays of the sun can kill bacteria and their spores in relatively short periods. As a rule, however, these experiments have been conducted in such a manner that the spores could germinate on the moist surface on which they were deposited (for instance, in Marshall Ward's experiments). That, however, does not at all conform to the conditions prevailing in planetary space. For Roux has shown that anthrax spores, which are readily killed by light when the air has access, remain alive when the air is excluded. Some spores do not suffer from insulation at all. That applies, for instance, according to Duclaux, to *Thyrothrix scaber*, which occurs in milk and which may live for a full month under the intense light of the sun. All the botanists that I have been able to consult are of the opinion that we can by no means assert with certainty that spores would be killed by the light rays in wandering through infinite space.

It may further be argued that the spores, in their journey through universal space, would be exposed during most of that period to an extreme cold which possibly they might not be able to endure. When the spores have passed the orbit of Neptune, their temperature will have sunk to $-220°$, and farther out it will sink still lower. In recent years experiments have been made in the Jenner Institute, in London, with spores of bacteria which were kept for twenty hours at a temperature of $-252°$ in liquid hydrogen. Their germinating power was not destroyed thereby.

Professor Macfadyen has, indeed, gone still further. He has demonstrated that micro-organisms may be kept in liquid air (at $-200°$) for six months without being deprived of their germinating power. According to what I was told on the occasion of my last visit to London, further experiments, continued for still longer periods, have only confirmed this observation.

There is nothing improbable in the idea that the germinating power should be preserved at lower temperatures for longer periods than at our ordinary temperatures. The loss of germinating power is no doubt due to some chemical process, and all chemical processes proceed at slower rates at lower temperatures than they do at higher. The vital functions are intensified in the ratio of $1 : 2.5$ when the temperature is raised by $10°$ C. ($18°$ F.). By the time that the spores reached the orbit of Neptune and their temperature had been lowered to $-220°$, their vital energy would, according to this ratio, react with one thousand millions less intensity than at $10°$. The germinating power of the spores would hence, at $-220°$, during the period of three million years, not be diminished to any greater degree than during one day at 10°. It is, therefore, not at all unreasonable to assert that the intense cold of space will act like a most effective preservative upon the seeds, and that they will in consequence be able to endure much longer journeys than we could assume if we judged from their behavior at ordinary temperatures.

It is similar with the drying effect which may be so injurious to plant life. In interplanetary space, which is devoid of atmosphere, absolute dryness prevails. An investigation by B. Schröber demonstrates that the green alga *Pleurococcus vulgaris*, which is so common on the trunks of trees, can be kept in absolute dryness (over concentrated sulphuric acid in a desiccator) for twenty weeks without being killed. Seeds and spores may last still longer in a dry atmosphere.

Now, the tension of water vapor decreases in nearly the same ratio as the speed of the reaction with lower temperatures. The evaporation of water—*i. e.*, the drying effect—may hence, at a temperature of $-220°$, not proceed further in three million years than it will in one day at 10°. We have thus several plausible reasons for concluding that spores which oppose an effective resistance to drying may well be carried from one planet to another and from one planetary system to another without sacrificing their vital energy.

The destructive effect of light is, according to the experiments of Roux, no doubt due to the fact that the rays of light call forth an oxidation by the intermediation of the surrounding air. This possibility is excluded in interplanetary space. Moreover, the radiation of the sun is nine hundred times fainter in the orbit of Neptune than in the orbit of the earth, and half-way to the nearest fixed star, Alpha Centauri, twenty million times feebler. Light, therefore, will not do much harm to the spores during their transference.

If, therefore, spores of the most minute organisms could escape from the earth, they might travel in all directions, and the whole universe might, so to say, be sown with them. But now comes the question: how can they escape from the earth against the effect of gravitation? Corpuscles of such small weight would naturally be carried away by any aerial current. A small rain-drop, $\frac{1}{50}$ mm. in diameter, falls, at ordinary air pressure, about 4 cm. per second. We can calculate from this observation that a bacteria spore 0.00016 mm. in diameter would only fall 83 m. in the course of a year. It is obvious that particles of this minuteness would be swept away by every air current they met until they reached the most diluted air of the highest strata. An air current of a velocity of 2 m. per second would take them to a height where the air pressure is only 0.001 mm.—*i.e.*, to a height of about 100 km. (60 miles). But the air currents can never push the particle outside of our atmosphere.

In order to raise the spores to still higher levels we must have recourse to other forces, and we know that electrical forces can help us out of almost any difficulty. At heights of 100 km. the phenomena of the radiating aurora take place. We believe that the auroræ are produced by the discharge of large quantities of negatively charged dust coming from the sun. If, therefore, the spore in question should take up negative electricity from the solar dust during an electric discharge, it may be driven out into the sea of ether by the repulsive charges of the other particles.

We suppose, now, that the electrical charges—like matter—cannot be subdivided without limit. We must finally come to a minimum charge, and this charge has been calculated at about $3.5.10^{-10}$ electrostatic unit.

We can, without difficulty, calculate the intensity of an electric field capable of urging the charged spore of 0.00016 mm. upward against the force of gravity. The required field-strength is only 200 volts per metre. Such fields are often observed on the surface of the earth with a clear sky, and they are, indeed, almost normal. The electric field of a region in which an auroral display takes place is probably much more intense, and would, without doubt, be of sufficient intensity to urge the small electrically charged spores which convection currents had carried up to these strata, farther out into space against the force of gravity.

It is thus probable that germs of the lowest organisms known to us are continually being carried away from the earth and the other planets upon which they exist. As seeds in general, so most of these spores, thus carried away, will no doubt meet death in the cold infinite space of the universe. Yet a small number of spores will fall on some other world, and may there be able to spread life if the conditions be suitable. In many cases conditions will not be suitable. Occasionally, however, the spores will fall on favorable soil. It may take one million or several millions of years from the age at which a planet could possibly begin to sustain life to the time when the first seed falls upon it and germinates, and when organic life is thus originated. This period is of little significance in comparison with the time during which life will afterwards flourish on the planet.'

The germs which in this way escape from the planets on which their ancestors had found abode, may either wander unobstructed through space, or they may, as we have indicated, reach outer planets, or planets moving about other suns, or they may meet with larger particles of dust rushing towards the sun.

We have spoken of the Zodiacal Light and that part of it which has been designated the counter-glow. This latter glow is regularly seen in the tropics and occasionally in that portion of our heavens which is just opposite the sun. Astronomers ascribe the counter-glow to streams of fine dust which are drawn towards the sun (compare page 147). Let us assume that a seed of the diameter of 0.00016 mm. strikes against a grain of dust which is a thousand times as large (0.0016 mm. diameter), and attaches itself to its surface. This spore will be carried by the grain of dust towards the sun; it will cross the orbits of the inner planets, and it may descend in their atmospheres. Those grains of dust do not, by any means, require very long spaces of time to pass from one planetary orbit to another. If we assume that the spore starts with zero velocity near Neptune (in which case the seed might originate from the moon of Neptune; for Neptune itself, like Uranus, Saturn, and Jupiter, is not yet sufficiently cooled to sustain life), the spore would reach the orbit of Uranus in twenty-one years, and of Mercury in twenty-nine years. With the same initial velocity such particles would be twelve years in passing between the orbits of Uranus and Saturn, four years between Saturn and Jupiter, two years between Jupiter and Mars, eighty-four days between Mars and the earth, forty days between the earth and Venus, and twenty-eight days between Venus and Mercury.

We see from these time estimates that the germs, together with the grains of dust to which they have attached themselves, might move towards the sun with much smaller velocity (from ten to twenty times smaller) without our having to fear any loss of their germinating powers during the transit. In other words, if these seeds adhere to the particles, ninety or ninety-five per cent. of whose weight is balanced by the radiation pressure, they may soon fall into the atmosphere of some inner planet with the moderate velocity of a few kilometres per second. It is easy to calculate that if such a particle should, in falling, be arrested in its motion after the first second, it would yet, thanks to the strong heat radiation from it, not be heated by more than 100° Cent. (212° F.) above the temperature of its surroundings. Such a temperature can be borne by the spores of bacteria without fatal effects for much more than one second. After the particles, together with the seed adhering to them, have once been stopped, they will slowly descend, or will be carried down to the surface of the nearest planet by descending convection currents.

In this way life would be transferred from one point of a planetary system, on which it had taken root, to other locations in the same planetary system which favor the development of life.

The seeds not caught by such particles of dust may be taken over to other solar systems, and finally be stopped by the radiation pressure of their suns. They cannot penetrate any farther than to spots at which the radiation pressure is as strong as at their starting-points. Consequently, germs from the earth, which is five times as near the sun as Jupiter, could approach another sun within a fifth of the distance at which germs from Jupiter would be stopped.

Somewhere near the suns, where the seeds are arrested by the radiation pressure to be turned back into space, there will evidently be accumulations of these seeds. The planets which circulate around their suns have therefore more chance of meeting them than if they were not in the vicinity of a sun. The germs will have lost the great velocity with which they wandered from one solar system to another, and they will not be heated so greatly in falling through the atmospheres of the planets which they meet.

The seeds which are turned back into space when coming near a sun will there perhaps meet with particles whose weight is somewhat greater than the repelling power of the radiation pressure. They would, therefore, turn back to the suns. Like the germs, and for similar reasons, these particles would consequently be concentrated about the sun. The small seeds have, therefore, a comparatively better chance of being arrested before their return to space by contact with such particles, and of being carried to the planets near that sun.

In this manner life may have been transplanted for eternal ages from solar system to solar system and from planet to planet of the same system. But as among the billions of grains of pollen which the wind carries away from a large tree—a fir-tree, for instance—only one may on an average give birth to a new tree, thus of the billions, or perhaps trillions, of germs which the radiation pressure drives out into space, only one may really bring life to a foreign planet on which life had not yet arisen, and become the originator of living beings on that planet.

Finally, we perceive that, according to this version of the theory of panspermia, all organic beings in the whole universe should be related to one another, and should consist of cells which are built up of carbon, hydrogen, oxygen, and nitrogen. The imagined existence of living beings in other worlds in whose constitution carbon is supposed to be replaced by silicon or titanium must be relegated to the realm of improbability. Life on other inhabited planets has probably developed along lines which are closely related to those of our earth, and this implies the conclusion that life must always recommence from its very lowest type, just as every individual, however highly developed it may be, has by itself passed through all the stages of evolution from the single cell upward.

All these conclusions are in beautiful harmony with the general properties characteristic of life on our earth. It cannot be denied that this interpretation of the theory of panspermia is distinguished by perfect consistency, which is the most important criterion of the probability of a cosmogonical theory.

There is little probability, though, of our ever being able to demonstrate the correctness of this view by an examination of seeds falling down upon our earth. For the number of germs which reach us from other worlds will be extremely limited—not more, perhaps, than a few within a year all over the earth's surface; and those, moreover, will presumably strongly resemble the single-cell spores with which the winds play in our atmosphere. It would be difficult, if not impossible, to prove the celestial origin of any such germs if they should be found contrary to our assumption.

THE END

www.ingramcontent.com/pod-product-compliance
Ingram Content Group UK Ltd.
Pitfield, Milton Keynes, MK11 3LW, UK
UKHW041951230426
12048UKWH00008B/266